THE CONTINUITY DEBATE:

Dedekind, Cantor,
du Bois-Reymond,
and Peirce on Continuity
and Infinitesimals

BENJAMIN LEE BUCKLEY

DOCENT PRESS
Boston, Massachusetts, USA
www.docentpress.com

Docent Press publishes books in the history of mathematics and computing about interesting people and intriguing ideas. The histories are told at many levels of detail and depth that can be explored at leisure by the general reader.

Cover design by Brenda Riddell, Graphic Details, Portsmouth, New Hampshire.

© Benjamin Lee Buckley 2008, 2012

All rights reserved. No part of this book may be reproduced or utilized in any form or by any means, electronic or mechanical, including photocopying and recording, or by any information storage and retrieval system, without permission in writing from the author.

For Hanne Blank – the first of many I shall write for you.

Acknowledgements

There are too many people to whom I owe my gratitude to be able to name them all here. There are, however, several who indeed must be singled out. I thus wish to thank the following people.

David Charles McCarty, who spent countless hours going over initial drafts and arguing proofs with me, but who also inspired me, supported me, taught me my history and philosophy of mathematics in the first place, and caught every last split infinitive I introduced into the text. (Any remaining split infinitives have been later additions).

Larry Moss, Frederick Schmitt, and Timothy O'Connor for their insightful comments and criticism. Particular thanks go to Dr. Moss for his mathematical feedback, and Dr. Schmitt for first suggesting the addition of C. S. Peirce to this work, thus inspiring my research to take on an entirely new direction from what it otherwise may have.

Alexander Hall, for assistance with the most recent drafts of this book, assistance both philosophical and grammatical, and for the lengthy philosophical discussions on many topics mathematical.

Hanne Blank, for assistance with proofreading and German spelling, for listening to me think out loud, for substantial general support and encouragement, and for letting me teach her how to count infinity.

Joe Decker, Erin Bolstad, Martin DeMello, Evan Wade Buckley, and, posthumously, Jean D'Amato Thomas, for the encouragement, intellectual conversation, and mathematical insight.

Christopher Raridan, Anthony Giovannitti, and the Clayton State University Math Club for giving me an opportunity to refine my analysis of Paul du Bois-Reymond's infinitesimal argument.

Christopher Keele, Timothy Keele, and Jacob Keele, for taking on much of the grunt-work associated with translating du Bois-Reymond.

And last but not least, my editor Scott Guthery, for his tireless efforts and multiple hours spent with this text in many versions. Any remaining errors are, of course, entirely my own.

Contents

List of Figures	ix
Chapter 1. Introduction	**1**
1.1. The Historical Importance of Mathematical Continuity	1
1.2. Infinity, Infinities, and Transfinites	3
1.3. Ghosts of Departed Quantities	6
1.4. Intuitive Continuity	8
1.5. Preliminary Remarks on Mathematical Continuity	11
1.6. Organization	15
Chapter 2. History of Continuity	**17**
2.1. Historical Context	17
2.2. Aristotle's Continuity	18
2.3. Archimedes and the Infinitely Small	22
2.4. The Medieval Debate on Motion, Change, and Continuity	24
2.5. Analysis	28
2.6. Conclusions	33
Chapter 3. Richard Dedekind	**37**
3.1. Biography and Introduction	37
3.2. Dedekind Cuts and the Principle of Continuity	39
3.3. Dedekind's Theory of Number	45
3.4. The Nature of Dedekind Continuity	52
3.5. The Relationship Between Dedekind's Continuity and Infinitesimals	56
Chapter 4. Georg Cantor	**61**
4.1. Biography and Introduction	61
4.2. Real Numbers (1872)	62

4.3.	Continuity and Denumerability (1872 - 1878)	67
4.4.	Early Real-Number Continuity (1878)	69
4.5.	Infinity, and the Definition of Continuity (1883)	72
4.6.	Infinitesimals (1883 and 1887)	76

Chapter 5. Paul du Bois-Reymond   83
  5.1. Biography and Introduction   83
  5.2. Infinitärcalcül   85
  5.3. Idealist versus Empiricist: Basic Theories, and Straight Lines 88
  5.4. Infinitesimals, For and Against   94
  5.5. Continuity and a Unified Mathematics   102

Chapter 6. Charles Sanders Peirce   107
  6.1. Introduction   107
  6.2. Synechism   108
  6.3. Early Definition of Continuity   110
  6.4. Middle Definition of Continuity   111
  6.5. Later Definition of Continuity   115
  6.6. Advantages and Disadvantages of Peirce's Late Continuity 121
  6.7. Peirce's Infinitesimals and his Mathematical Continuity   123
  6.8. Peirce's Infinitesimals and his Final Definition of Continuity 128

Chapter 7. Infinitesimal Interlude   131

Chapter 8. Conclusions   139
  8.1. Introduction   139
  8.2. Summary of Our Four Figures   140
  8.3. Compositional Continua and Aristotle   149
  8.4. Continuity, Cantor, and Dedekind   152
  8.5. Continuity and Peirce   157
  8.6. Continuity and du Bois-Reymond   159
  8.7. Conclusion   162

Bibliography   165

List of Figures

1.1	Equilateral Right Triangle	13
1.2	Measurement of Triangle	13
2.1	Method of Exhaustion, 1st Inscription	23
2.2	Method of Exhaustion, 2nd Inscription	23
7.1	Circle with Tangent	132
7.2	Curvilinear Approximation, Step 1	132
7.3	Curvilinear Approximation, Steps 2 through n.	133

CHAPTER 1

Introduction

1.1. The Historical Importance of Mathematical Continuity

In the last century, differential and integral calculus have been commonly referred to as "the calculus," in much the same way as medieval scholars referred to Aristotle as "the Philosopher" and ancient Romans referred to Virgil as "the Poet." It is as though there is no other, or at least no better, way of calculating. It was not always so. Calculus was born in conflict, with Gottfried Wilhelm von Leibniz (1646–1716) and Sir Isaac Newton (1643–1727) both staking the claim of invention and initiating a vehement argument about the philosophical details underpinning their mathematical procedures. Calculus was invented in the seventeenth century, but did not gain prominence in the mathematical world until nearly two hundred years later, early in the nineteenth century. The delay was largely because these same philosophical differences had not yet been settled, or even directly addressed.

The main philosophical objection to early acceptance of the calculus surrounded the mysterious infinitely small parts of the equations. Calculus, as a means of calculating motion at an instant, necessitates some method of referring to the distance between two numbers which are infinitely close. Leibniz called this distance an *infinitesimal*, but claimed the infinitesimal was only a useful fiction, not an actual number. Newton referred to this small distance sometimes as a *fluxion* – the rate of change of flowing quantities over time – and sometimes as an infinitesimal. While both mathematicians hinted at the possibility of something like limit theory taking the place of fluxions or infinitesimals, limit theory was not mathematically formalized until the early nineteenth century.

Limit theory, roughly, is the mathematical system that uses variables to allow one quantity to approach infinitely closely to another but never reach it. Thus, limits seemingly rid mathematics of the need for infinitely small entities: numbers no longer have to reach infinitely small distances, as the gap between them (at all times finite) could decrease infinitely. The introduction of this theory allowed calculus to move forward and gain wider acceptance, but it was not itself universally accepted, some claiming it raised more philosophical difficulties than it solved. In particular, it was limit theory which necessitated that the real numbers themselves display a kind of continuity. If calculus was to be done algebraically, without integral recourse to geometry, the numbers themselves must be able to make the continuous approach necessitated by limits, and thus, the continuity of the real numbers must first be established.

The topic of this book is the historical struggle to define and defend a real-number continuum which could do the work limit theory required of it. These definitions drew heavily on philosophical and foundational assumptions, and each raises numerous philosophical questions of its own. As we shall see, attempts to formulate a non-geometrical mathematical continuity raise questions such as: What is a number? What, in particular, is a real number? What is the true nature of continuity itself? Does a philosophically coherent definition of continuity logically commit us to infinitesimally small quantities? Is the concept of an infinitesimally small quantity even logically coherent? What is the relationship between this real number continuum and other well known continua, such as the geometrical straight line? The main question to be addressed, of course, is whether mathematical continuity exists at all.

For most of the twentieth century, it was largely assumed that these questions were settled, as far as calculus went. In the argument between mathematicians calculating with infinitesimals and those calculating with limits, the limit theorists were believed to have won. The infinitesimal theorists, it seemed, had failed to present a coherent mathematical infinitesimal calculus, and meanwhile, limit theorists had grand successes. However, in the 1960s, German-born mathematician Abraham Robinson (1918–1974) presented a mathematically coherent and powerful calculus which incorporated infinitesimals, transfinite quantities,[1]

---

[1] Just as an infinitesimal is an actually infinitely small quantity, a transfinite is an actually infinitely large quantity. As we shall see below, Georg Cantor proved that there were infinitely many different magnitudes of infinity, and thus, they can be ordered and used in calculations.

and real numbers into a single system, thereby reawakening the sleeping dragons of the previous century. It is my hope that by reexamining the fertile mathematical and philosophical arguments of the past we can better frame the current debate about the philosophical underpinnings of the standard calculus and Abraham Robinson's non-standard analysis. As such, this book focuses on the debate surrounding the nature of real numbers and whether they form a continuous entity, as it occurred in the late nineteenth century, as calculus became more widely accepted. The figures under consideration all had one foot in mathematics, and one foot in philosophy: German mathematicians Richard Dedekind, Georg Cantor, and Paul du Bois-Reymond; and American philosopher Charles Sanders Peirce.

The purpose of this introductory chapter then is to define and discuss in a preliminary fashion certain concepts which shall be referred to often throughout the course of our considerations, concepts relied upon in our investigation of continuity: infinity, infinitesimal quantities, and continuity itself. More precise and philosophical discussions of continuity will be presented in subsequent chapters; this chapter will focus instead on intuitive notions. However, we will follow our discussion of intuitive continuity with some preliminary remarks on the concept of mathematical continuity. The chapter will end with an outline of the structure of the remainder of the book.

1.2. Infinity, Infinities, and Transfinites

Infinity has had a long philosophical history. Some have viewed actual infinity as simply impossible, others have viewed it as simply factual, and still others as mystical, mysterious, or even divine. For example, one of René Descartes' (1596–1650) arguments for the existence of God depends intimately on the concept of the infinite:

> So from what has been said it must be concluded that God necessarily exists. It is true that I have the idea of substance in me in virtue of the fact that I am a substance; but this would not account for my having the idea of an infinite substance,

> when I am finite, unless this idea proceeded from substance which really was infinite.[2]

The infinite, for Descartes, is too large to be comprehended without the help of a similarly infinite being – thus, without the existence of God, humans could never comprehend the infinite. About twenty years after Descartes wrote his *Meditations,* Baruch Spinoza (1632–1677) used the infinite to characterize his concept of God as well. Proposition XXI of Spinoza's *Ethics* is,

> ... all things which follow from the absolute nature of any attribute of God must always exist and be infinite, or, in other words, are eternal and infinite through the said attribute.[3]

Thus, according to Spinoza, it is not only necessary that God be infinite, but also that everything that comes from God must be similarly infinite.

Not all discussions of the infinite are theological, of course. Aristotle (384–322 BCE) considered the infinite in Book I of the *Physics*. When Aristotle laid out the principles of the physical world, he wished to begin by discussing the most basic concepts, and for him the most basic concept is that of number. When considering a thing, that thing must either be one or more than one. If one, it must either be still or in motion; if more than one, it must be either finite or infinite. Thus, number, motion, and the infinite must be addressed before anything else in the physical world.[4]

Aristotle distinguished between potential and actual infinity. Potential infinity is infinite in the sense that one can always find or create a bigger number, another instant, a further corner of space, but at any given time, only finitely many such objects are considered. Actual infinity is a collection of which it could be said that every one of infinitely many objects exist, simultaneously. Aristotle himself believed that potential infinity was a more coherent concept, and that actual infinity was problematic. He thus argued that space is infinitely divisible, but only potentially so: we could always perform another division on a

---

[2] See Descartes [1996, p. 31].
[3] See Spinoza [1989, p. 58].
[4] Aristotle, *Physics*, Book I, 184b15–22. Throughout the dissertation, all Aristotle quotations refer to Barnes [1984]. The pagination is the standard Aristotelian pagination, however.

spatial entity, no matter how many we have already performed, but this infinite division could never be completed, one could never have an actually infinitely divided section of space. For Aristotle, the infinite, in division, only exists potentially.[5]

As for infinite magnitude, Aristotle argued that there is no such thing. Infinitely many things do not exist in actuality, only finitely many. Infinite magnitude does not even exist potentially; for "it is impossible to exceed every definite magnitude, for if it were possible there would be something bigger than the heavens."[6] However, Aristotle did not want mathematicians to despair over the non-existence of the actual infinite, or the lack of infinite magnitude. "In point of fact, they do not need the infinite and do not use it."[7] Thus, mathematicians would not miss infinite magnitude, and they have as much infinite division as they need.

Georg Cantor (1845–1918), whose theory of continuity is discussed in Chapter 4, would disagree. His transfinite theory pushed the idea of the infinite to new levels. Cantor not only argued for the mathematical existence of actual infinities, but he went on to establish different magnitudes of infinity, and began to calculate with them.[8] Transfinite theory, the theory that these different magnitudes of infinity can be specified with precision and then well ordered, is possibly the most controversial theory Cantor forwarded, but classical set theory, his most popular, also relies on the idea of an actual infinite.

Richard Dedekind (1831–1916), the subject of Chapter 3, also played an important historical role in the development of the actual infinite, by presenting a solution to a long standing paradox of infinity. Ancient and medieval scholars puzzled over what they termed the "paradox of unequal infinities": the fact that in some instances, a part of an infinite set is the same size as the infinite set itself. For example, consider the set of natural numbers, beginning with

---

[5]Ibid., Book III, 207b10–15.
[6]Ibid., Book III, 207b20–21.
[7]Ibid., Book III, 207b30.
[8]See in particular [Cantor, 1955]. It should be noted that Cantor was not the first to argue the existence of the actual infinite, nor was he the first to speculate that there was more than one type of infinity. He was, to my knowledge, the first who not only proved more than one magnitude of infinity, but who also proved the existence of infinitely many magnitudes of infinity.

one, and consider the set of only the even numbers. It seems as though there should be half as many even numbers as natural numbers; and yet, every even number corresponds to exactly one natural number, and vice versa. Thus, the evens form a proper part of the naturals, while being equinumerous with the supposed larger set. This violates our intuitions that a part of a whole must be smaller than the whole itself.[9] Dedekind stipulated that rather than being paradoxical, this is the very definition of infinity. "A system S is said to be *infinite* when it is similar to a proper part of itself; in the contrary case S is said to be a *finite* system."[10] Thus, the fact that the even numbers can be put into one-to-one correspondence with the set of all natural numbers is not a paradox, but rather proof that the natural numbers are indeed infinite. This property does not hold of finite sets. We will see to what use Dedekind puts this definition of the infinite in Chapter 3.

1.3. Ghosts of Departed Quantities

The infinitely small is no less vexing or inspiring than the infinitely large. A perfect example of the vexing nature of the infinitely small is given by George Berkeley (1685–1753) when he suggested that Newton's fluxions be defined as, "ghosts of departed quantities."[11] Though Berkeley was speaking of fluxions in particular, his jibe seems to apply equally well to infinitesimal quantities and to limits. In all these cases, we are attempting to capture a measurement of a magnitude as its variable determinant shrinks beyond all finite bounds. Calculus seems to require us to perform calculations using a quantity that is smaller than any given quantity. Limits are not claimed to be quantities, and thus avoid these particular difficulties, though we are left with the question of what, exactly, a limit is.

Though calculus brings the infinitesimal debate into the realm of mathematics, the question of the infinitely small predates Newton and Leibniz, appearing, for example, in theological questions such as the old medieval chestnut, "How many angels can dance on the head of a pin?" The question is almost a joke

---

[9] St. Bonaventure (1221–1274) and al-Ghazali (1058–1111) used this paradox as part of an argument against the eternity of celestial motion, for example.

[10] See Dedekind [1963, p. 63]. C.S. Peirce, as we will see in Chapter 6, had a similar proof of infinity, which involved what he called the "inference of transposed quantity."

[11] See Berkeley [1901, Paragraph XXXV].

1.3. GHOSTS OF DEPARTED QUANTITIES

today, but seven hundred years ago, determining the answer to the question took some thought, as angels were thought to have no physical extension at all. The question, then, of how many things lacking physical extension can fit in a particular space was a controversial one, and mathematics contains an updated version of this puzzle when we ask whether it is possible for a line to be composed of points. A similar medieval question wondered how angels could move through space – whether it was logically possible for something non-extended to move through extended space.[12]

Though philosophically troublesome, infinitesimal quantities are held by some to be necessary elements of anything continuous. Paul du Bois-Reymond (1831–1889), the subject of Chapter 5, believed infinitesimals to be necessary for continuity, and thus, necessary not only for calculus, but for geometry and for any other branch of mathematics or science that dealt with the continuous. Because points are non-extended non-spatial entities, du Bois-Reymond was convinced that a straight line could not be composed merely of points. Continuity required that the line contain intervals of infinitely small length, that is, infinitesimal intervals, in addition to points.

Charles Sanders Peirce (1839–1914), the subject of Chapter 6, came to a similar belief late in his career – the belief that points were insufficient for continuity, but also that numbers were insufficient for similar reasons. The solution Peirce gave is quite different than du Bois-Reymond's, but Peirce also believed in the importance of infinitesimal quantities, and developed an infinitesimal theory in stages throughout his philosophical career. Unlike du Bois-Reymond, Peirce never argued that infinitesimal quantities followed logically from our assumptions of continuity. He did, however, argue that they were necessary for a proper understanding of continuity itself, and infinitesimals become necessary features of calculation with his final and most complex theory of continuity. While du Bois-Reymond argued that the real number system of Cantor and Dedekind was inadequate, Peirce agreed but went one step further, arguing that limit theory itself was philosophically confused. Both men believed that infinitesimals helped solve these mathematical and metaphysical problems.

---

[12]See, for example, the first quodlibet of William of Ockham, question 5, "Can an angel move locally?" in [Ockham, 1991]. The puzzle was briefly this: to say that x moved is equivalent to saying that x was here, and is now there (with some important codicils). An angel, as a non-extended, non-spatial entity, can never occupy a position in space, and therefore can never *be* here or there for any particular space.

8 1. INTRODUCTION

Thus, the concept of the infinitesimal is one which will recur in every chapter in this book. Both Dedekind and Cantor believed infinitesimals were either impossible or at least philosophically suspect, which is one possible motivation behind their attempts to solve the philosophical difficulties with limit theory, by shoring up the continuity of the set of all reals. Du Bois-Reymond and Peirce believed infinitesimals were necessary parts of any actual continuity, and thus, should not be abandoned in our systems of calculus; both disagreed strongly with the anti-infinitesimal systems of continuity constructed by Cantor and Dedekind.

1.4. Intuitive Continuity

A necessary first step in a detailed investigation of the metaphysics and epistemology of mathematical continuity is to define continuity itself. Often, mathematicians appeal to our intuitions about continuity when forwarding a rigorous definition of continuity. Peirce, for example, thought that mathematical definitions of continuity must live up to our philosophical intuitions, and Dedekind held that the true nature of continuity could be discovered by examining our intuitions of what makes a straight line a continuum. To attempt to examine these intuitions about continuity, we will briefly look at the discussion of a distinctly non-mathematical continuum – that of the color-wheel – in Friedrich Waismann's *Introduction to Mathematical Thinking*, [Waismann, 2003].

Waismann asked, when you look at the color wheel, one which does not segment into, say, six distinct colors, but rather represents each color as it blends into the next, how many colors do you see? Does each color nuance represent a new color? If it does, and the color spectrum is continuous, then there should be infinitely many colors in a spectrum. But do you see infinitely many colors? Waismann answers "No."

> The color continuum has a structure entirely different from that of the number continuum. In the case of two real numbers it is uniquely established whether they are equal or different. No matter now close they may lie near one another on the number axis, they are and always will be different numbers. A

color, however, runs into another imperceptibly, it blends with it; more accurately stated, it is without meaning to speak of isolated elements out of which the continuum is to be erected. The concept of number is not applicable to formations of this kind, for the first assumption of counting is that the entities to be counted be clearly distinguishable.[13]

Thus, colors are indistinct from one another, blending and merging in ways that are predictable but not always perceptible. The color wheel, as a representation of this color continuum, thus differs from crayons in a box, each of which represents a distinct, identifiable color.

Though Waismann himself does not call the color wheel the epitome of intuitive continuity (he discusses intuitive continuity directly, though he does not define it), the changes on the color wheel do have features most people would associate with continuity on an intuitive level. The lack of any clear distinction from one place on the wheel to the next, the merging of one part into another, and the lack of gaps or spaces, all seem important features of continuity. For all their plurality, the real numbers, the nature of which we shall consider in detail in the following chapters, are in some sense distinct. They are a totally ordered set, and for any two non-identical real numbers, we can distinguish them from each other – they do not blend into one another. Thus, the main question with respect to the continuity of the real numbers is whether it is possible for a continuum to be composed of distinct objects.

Typically, we view distinct objects, or even collections of distinct objects, as the contrary of continuity. To continue with our investigation of our intuitions on the matter, consider objects which are clearly non-continuous, such as such as tables, or a glass of water. A table is not continuous, no matter how smooth, flat, or large it might seem; once it is divided enough times, it becomes something different. Some tables, chopped in two, might still function as tables, but chopped again and again, they cease to be tables, and instead become firewood. Water will take finer-grained divisions before you get something non-watery, such as individual hydrogen and oxygen atoms, but only a finite number of divisions is needed to reach that level; water is no more continuous, at bottom, than a table.

---

[13]Ibid., p. 212–213.

We may argue, drawing on our above examples, that a color wheel is not absolutely continuous either; color, after all, is a physical thing. We may take smaller and smaller slices of our color wheel without seeing any breakdown of continuity; but that is more a function of the limits of our eyes than it is true indication of the continuity of the color wheel. If we could focus on a small enough wedge of the color wheel, we would instead see a collection of atoms or sub-atoms, which are too small to reflect light and therefore colorless. Our allegedly continuous color wheel breaks down. This raises two important questions. First, is anything at all actually continuous? And second, if the answer to the first question is yes, what would the nature of this actually continuous entity be?

Space, time, and motion have been considered actually continuous at least since Aristotle. There are, however, those who hold that some or all of the above are non-continuous – that time, for example, is composed of discrete atomic instants – but if continuity ever exists in the physical world, these are the three top contenders for continuous entities.[14] Indeed, it seems as though no matter how short a segment of space you have, it will resemble any other segment of space in important respects; you can divide space into smaller and smaller parts and never reach anything qualifying as "non-space."[15] Also, one part of space merges into the next without identifiable borders, distinctions, or gaps. Time, too, at least appears to flow by smoothly, without bits of non-time breaking things up. Motion is more complex, usually analyzed as an interaction between space and time, and therefore in some sense continuous; unlike space and time, however, motion is usually limited to particular events, and even particular objects. In Chapter 2, the continuity of space, time, and motion is discussed in further detail.

---

[14]As an interesting side note, Immanuel Kant (1724–1804) did not believe that space or time were part of the physical world, but rather, that they existed as the very conditions of our perception of the external world. However, he still held that they were continuous entities. (See Kant [1965] A26/B42, and A33. See also A213/B260 and A170/B212 for Kant's discussion of the continuity of space and time, respectively.) This seems to suggest that for Kant, continuity itself was a condition of perception, rather than a property of the external world.

[15]Interestingly, however, some modern physicists are questioning the continuity of space, arguing that a Planck length is the smallest possible unit and that space is composed of these.

1.5. Preliminary Remarks on Mathematical Continuity

While continuity has been an essential part of geometry at least since Euclid (325–265 BCE), numbers have often been seen as the antithesis of continuity. Aristotle and other ancient Greeks viewed geometry as the science of space (and thus naturally continuous) and mathematics as the science of number. The concept of number, for Aristotle, began with 1, and the collection of numbers included only multiples of 1. As Aristotle believed in potential, but not actual, infinity, there were as many numbers as you wanted, but at any one time you only had finitely many of them, and numbers were in no sense infinitely divisible; you could only divide numbers until you returned to the smallest number: the unit. The set of natural numbers is clearly non-continuous; each one is uniquely identifiable, there is no smooth connection between them, and they are discrete at least in the sense of being fully distinguishable, one from the other.

As we will see in Chapter 2, during the sixteenth and seventeenth centuries, the concept of number broke out of this Aristotelian mold, and the division between mathematics and geometry began to blur. Number systems grew to include the negatives numbers, the rationals, and the the irrationals. Descartes's advances in algebra and geometry allowed geometrical shapes to be described completely numerically. Thus, as numerical and mathematical systems grew, it became at least possible to view numbers as in some sense continuous.

While it is clearly impossible to view the natural numbers as continuous, the same difficulties do not automatically apply to the set of rational numbers. The rational numbers are distinct in the sense of being uniquely identifiable and distinguishable one from the other. However, the rationals have a feature the naturals lack. The rationals are dense: between any two rational numbers, there exists another rational. Furthermore, there exist infinitely many rationals between any two. Though $2/3$ does not run and bleed into $3/4$ the way red and orange subtly merge, it can be said that $2/3$ and $3/4$ are parts of a spectrum of their own sort, with the infinitely many rationals between them resembling the infinitely many colors of the color wheel. Rational numbers are distinguishable one from the other, but there is no *next* rational number – in this sense, the rationals could be said to be smooth.

However, the rational numbers do not form a continuous set. The reason for this failure highlights a third feature of our intuitions about continuity (the first two being non-discreteness and smoothness): something which contains gaps is not continuous. A span of time from which an hour has been removed is no longer a continuous span. The time previous to this missing hour can be continuous, as can the time after – continuity does not require all time to be present at once – but the time span with the gap in the middle loses its cohesion, and therefore its continuity. The rational numbers have identifiable gaps; there are lengths which no rational number can measure.

Bertrand Russell (1872–1970) discussed these gaps in his *Principles of Mathematics*:

> In the last chapter of part III, we agreed provisionally to call a series continuous if it had a term between any two. [...However,] ever since the discovery of incommensurables in Geometry – a discovery of which is the proof set forth in the tenth Book of Euclid – it was probable that space had continuity of a higher order than that of the rational numbers.[16]

Russell believed that density was a type of continuity; yet, due to the incommensurability (i.e., irrationality, a number which cannot be expressed as a ratio between whole numbers) of certain geometrical measurements, space is richer, more complete, than the rational numbers. This incommensurability is easy to demonstrate. Consider a right triangle with two sides of length 1 in Figure 1.1.

Now take the three sides of the triangle and lay them end to end along a straight line, comparing them to the rational number line as shows in in Figure 1.2.

The top line in Figure 1.2 is our deconstructed triangle; the bottom line is the rational number line, attempting to measure the line above. *A* clearly corresponds to 1, *A* and *B* together are 2. However, our rational numbers contain no number corresponding exactly to the sum of all three sides. We could find a rational that estimated the measurement of this line with some precision. The value 3.4, for example, would be quite close, and 3.41 would be even closer.

---

[16]See Russell [1903, p. 287].

1.5. PRELIMINARY REMARKS ON MATHEMATICAL CONTINUITY

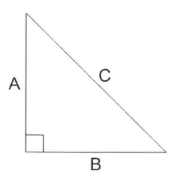

FIGURE 1.1. Equilateral Right Triangle

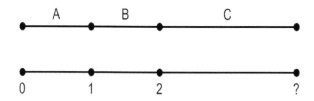

FIGURE 1.2. Measurement of Triangle

However, the exact measurement of this line is beyond the capability of the rational numbers. The rationals have gaps when we attempt to compare them to geometrical structures. An entity with obvious holes violates our intuitive sense of continuity.

The real numbers, unlike the rationals, are capable of measuring these incommensurable geometrical forms. With the reals, we no longer find gaps of this kind when we attempt to match our number line against the geometrical line, as our real number system captures them all. Were the second line in Figure 1.2 the real number line rather than the rational, there would be a number, namely, 2 plus the square root of 2, precisely where the last point occurs; there is no gap here.

Dedekind will argue convincingly that there are no gaps at all in the real number system.[17] However, while the real numbers have the property of density, just as the rationals do, and while they have no obvious holes, there is still some sense in which the set of real numbers is composed of discrete entities. While it is harder to specify the irrational real numbers with exactitude, the real numbers are still discrete in the following sense: given any two non-identical real numbers, it is possible to distinguish them, one from the other. 2 and $\sqrt{2}$ are easily identified and unique, and distinguishable, but so are $\sqrt{2}$ and 1.4142, though they are closer to each other.

As we will see in Chapter 2, Aristotle argued that a continuum could not be composed of discrete entities, therefore, to the extent that $\sqrt{2}$ and the rest of the real numbers are viewed as discrete entities, Aristotle would argue that the set of reals is not continuous. These two issues – whether real numbers are discrete entities, and whether continua can be composed of discrete entities – will become essential in much of what follows. Dedekind did not address this difficulty directly, but followers of Dedekind have taken various approaches, some arguing that Dedekind's real numbers are not discrete entities, some arguing that Dedekind's continuum proves that continua can in fact be composed of discrete objects. Cantor called the "point-continuum" (that is, a continuum composed of discrete entities such as points or numbers) the very essence of continuity. Du Bois-Reymond and Peirce believed that numbers (and points on a line, for that matter) are indeed discrete entities, and they followed Aristotle in asserting that continua cannot be composed of discrete entities. Thus, whether a continuous entity can be composed of discrete elements is the very issue around which much of our considerations turn.

---

[17]Of course, this does not mean the set of real numbers is absolutely complete, in the sense of containing every possible number or being able to perform every possible mathematical calculation; the reals do not contain the imaginary numbers, for example, nor do they contain transfinites. They are gap-free in a very specific sense, which will be discussed more thoroughly in later chapters. As we shall see in Chapter 6, the fact that the set of real numbers does not contain every possible number was significant in Peirce's determination that this set was not continuous.

1.6. Organization

The debate over the discrete versus the continuous is just one of many issues which will be examined carefully in the chapters that follow. The real number continuum and its development in the late nineteenth and early twentieth centuries, as we shall see, is rich with philosophical ramifications of all sorts. In order to give a background to the debates of these four figures, as well as to highlight the philosophical concerns which have followed continuity throughout history, Chapter 2 shall follow the history of theories of continuity from Aristotle to Abraham Robinson, in brief outline. After setting this ground, the next four chapters present in-depth analyses of our four figures in roughly historical order: Dedekind, Cantor, du Bois-Reymond, and Peirce.

Dedekind's ground-breaking essay, "Continuity and Irrational Numbers," which explains the Dedekind-cut theory of real numbers, will be discussed in detail in Chapter 3, as will Dedekind's philosophical view of the nature of numbers. Chapter 3 also demonstrates that Dedekind's continuity is mathematically inconsistent with the existence of infinitesimal quantities. Chapter 4 presents Cantor's theory of continuity as he developed it over time, beginning with his formulation of real numbers in 1872 and ending with his specification of necessary and sufficient conditions for continuity from his *Gründlagen* of 1883. Chapter 4 also treats Cantor's argument that infinitesimals are self-contradictory and therefore impossible.

In Chapter 5, we move toward mathematicians who believed infinitesimals are consistent with continua, and consider du Bois-Reymond's real number theory, theory of continuity, and infinitesimal theory. Chapter 5 is mainly concerned with du Bois-Reymond's *General Theory of Functions*, in which he wished to thoroughly investigate the philosophical groundwork and mathematical intuitions for functions themselves. This chapter analyzes in detail du Bois-Reymond's theory of continuity as well as his argument for the necessity of infinitesimal magnitudes.

Chapter 6 is dedicated to Peirce's work. Like Cantor, Peirce developed his theory of continuity over time; unlike Cantor, Peirce changed his view wildly. Chapter 6 thus follows this development in Peirce and analyzes the ramifications of each of his theories of continuity. This chapter also contains Peirce's criticisms of Cantor and Dedekind, and considers the interesting relationship between his theory of infinitesimals and his theories of continuity.

Chapter 7 is a short chapter meant to defend the idea that infinitesimals are not *prima facie* useless as mathematical entities, thus attempting to overcome a frequent criticism of infinitesimals; i.e., the charge that even if they were consistently definable, they serve no mathematical purpose. The chapter presents a few mathematical examples in which the use of infinitesimals seems to add in a productive way to the calculation.

Chapter 8, the concluding chapter, begins with a summary of the four theories of continuity discussed in this book, and presents an analysis of each in turn. Following this analysis, Chapter 8 will conclude that Cantor and Dedekind, by allowing the composition of continua from discrete elements, create problematic theories. Also, while du Bois-Reymond and Peirce forward fair criticisms of Cantor and Dedekind, ultimately, Peirce's most mature theory of continuity is even more problematic, leaving du Bois-Reymond's as the most promising theory of the group. Finally, Chapter 8 indicates how the results of this analysis might be applied to the philosophical investigation of contemporary theories of infinitesimals.

CHAPTER 2

History of Continuity

2.1. Historical Context

As was noted in Chapter 1, the concept of mathematical continuity is tied intricately to particular mathematical developments and systems, and thus to a particular point in history. Mathematical continuity is defined as continuity as it is meant to apply to systems of numbers, and thus only becomes possible with the introduction of real number theory in the sixteenth century, and only becomes relevant to calculus with the formal introduction of limit theory in the early nineteenth century. However, just as intuitive notions of continuity helped give our preliminary discussion of mathematical continuity a framework for discussion, so too an overview of philosophical approaches to the concept of continuity will prove highly useful in analyzing mathematical continuity throughout the rest of this book. Toward that end, this chapter presents a brief overview of the main philosophical approaches to continuity through the ages, as well as an overview of some of the historical mathematical developments most relevant to this project.

We begin with ancient Greece. Section 2.2 briefly considers some of Aristotle's more important contributions to our understanding of the concept of continuity, especially his definitions of continuity itself, and his argument that continua cannot be composed of indivisible entities. Section 2.3 discusses Archimedes, whose startlingly modern methods of geometrical calculation led him to a concept of continuity which differed starkly from Aristotle's. Section 2.4 raises some of the more interesting medieval debates concerning continua, especially surrounding the discussion of how one ought to understand motion and change. These debates on motion lead naturally to the early-modern period, discussed in Section 2.5; this section focuses on the development of the calculus – one goal of which is to be able to measure motion at an instant. Thus, Section 2.5 glances at the mathematically fertile century before calculus itself was invented, then looks into how the idea of continuity was central to the ensuing controversy which surrounded the calculus for two centuries after its introduction.

2.2. Aristotle's Continuity

Continuity was an important concept for Aristotle, and he spent much of the *Physics* considering its various properties. Aristotle wished to understand continuity thoroughly in order to understand space, time, and motion, all of which he saw as necessarily continuous entities; he also wished to develop a theory of continuity which could be used to overcome the paradoxes of Zeno (c. 490–430 BCE). In this section, however, we shall side-step Aristotle's actual applications of his theory in favor of a look into the theory itself, particularly focusing on his argument that a continuum can not be composed of indivisible entities. This argument, as well as the distinctions and definitions he uses to develop the argument, will be directly relevant to material in later chapters.

Aristotle explicitly equated continuity with infinite divisibility. Thus, for Aristotle, "the continuous is divisible *ad infinitum*"[1] and also the reverse, "what is infinitely divisible is continuous."[2] Similar statements appear throughout the *Physics*. As we saw in Chapter 1, our intuitions on continuity agree that entities which are destroyed by multiple divisions do not seem continuous to us. If we divide a unit of space, the result is still a unit of space; if we divide a table, the result is firewood. Tables are not, therefore, continuous.

However, by emphasizing *infinite* divisibility, Aristotle was not simply invoking the idea that space was not destroyed by division. He went beyond the simple intuition expressed above when he elaborated further: "By continuous I mean that which is divisible into divisibles that are always divisible."[3]

---

[1] See Aristotle, *Physics*, Book I, 185b10.
[2] Ibid., Book III, 200b18.
[3] Ibid., Book VI, 232b23.

In other words, Aristotle believed there can be no end to the division of a continuous entity - no smallest part can be reached through division. This is also an early example of what we will, later on in this work, call the 'mirror property' of continuity; there is at least some sense here that the parts of a continuum must have some qualitative resemblance to the continuum as a whole. Here, that resemblance exists in the form of continued divisibility. Thus, a point cannot be a part of the continuum. This does not imply that the line contains no points, only that successive divisions of the line never produce points or anything else non-divisible; the line cannot be decomposed into points.

Neither, for Aristotle, can a line be composed of points.[4] Aristotle's argument for the non-compositionality of the line relies on a three-part distinction: things "next to" each other are either in succession, contiguous, or continuous. Note that this is an inclusive and not an exclusive disjunction, as contiguous things are also in succession, and continuous things are contiguous – the distinction is rather a matter of degree of closeness. Also note that Aristotle means this to be an exhaustive definition; these are the only three ways in which two things can be said to be next to each other.

Briefly, Aristotle's tripartite distinction is this: two things are *in succession* if one immediately follows the other without something of the same kind in between. There could be something of a different sort between them, such as air between two people standing in succession in a line at the movies, but there is no third person between them. Two things are *contiguous* if they are in succession, and the outer extremity of one touches the outer extremity of the other. Two books next to each other on a shelf are contiguous if their covers are in physical contact. They are in succession because there is no third book between them; they are contiguous because their covers physically touch each other. Finally, two things are *continuous* if the outer extremities, rather than merely touching, are in fact one – continuous things share an outer extremity. Two countries which share a land mass, such as Egypt and Sudan, form a continuous entity: unlike books with separate (but physically touching) covers, there is only one border between Egypt and Sudan; the southern border of Egypt is also the northern border of Sudan, they thus share an outer extremity.

---

[4]Ibid., Book VI, 232a4.

With this distinction in mind, we turn now to Aristotle's main argument for the conclusion that continua cannot be composed of indivisibles.[5] Assume for *reductio* that there is a continuum composed exclusively of indivisibles, such as points.[6] It follows, then, that there must exist at least two points which are next to each other – for if there were not, then either there would be a space between them and thus the continuum would not be continuous, or there would be some non-point thing in between them, and thus the continuum would not be composed exclusively of points. Consider now, in what sense these two points could be next to each other, given our previous distinction. They cannot be merely in succession, as that would allow for gaps which would destroy the continuity. Thus they are either contiguous or continuous. However, contiguity requires contiguous entities to have extremities which touch each other, and points, being indivisible and thus without parts, have no such extremities.[7]

Thus, there is only one option left; the two points must be next to each other in a continuous manner, i.e., they must touch such that their extremities are one. However, as we have just noted, indivisible entities have no extremities; in fact, since points have no parts at all, no part of a point can touch a part of another point. The only manner in which these neighboring points could be next to each other is to touch each other by being in exactly the same spot as one another; i.e., they must be "in contact with one another as whole with whole."[8] However, if this were so, the entire continuum would be the extension of a single point, which is impossible, as continua are infinitely divisible, and points are indivisible. As we have exhausted all of the possible ways in which two things can be next to each other, we must conclude that two points on a line cannot be next to each other in any way. As our assumption that there was a continuum composed exclusively of points led to the conclusion that two points on the line must be next to each other, we must conclude that continua *cannot* be composed of points.

---

[5]Ibid., Book VI, 231a18–231b16.

[6]Aristotle states the argument in terms of points, but notes that the same reasoning applies to any type of indivisible.

[7]Just to make the reasoning here explicit: if a point had an outer extremity with which it could touch a bordering point, then that point would be divisible into outer extremities and inner non-extremities – and would therefore be divisible.

[8]Ibid., Book VI, 231b2-3.

The very first step of this argument is one which would not be accepted by those who view the real numbers as continuous; it also would not be accepted by those who view the geometrical line as composed of points. The first step, recall, is the claim that if a continuum is composed of points, it follows that two points must be next to each other. Those who have some experience with the real numbers, or even the rational numbers, will recognize that this contradicts the property of density: between any two real numbers, there is always a third, and the same holds true of the rationals. Density is also a property of the geometrical line: between any two points, there is another point. This third point does not constitute a gap in the continuum, and it does not constitute a non-point entity, therefore, this third alternative has not been ruled out by Aristotle. In other words, two points cannot be next to each other, but what intervenes between them is nothing which disrupts continuity, because what intervenes between them is simply more points. The ramifications of this third alternative will be discussed in later chapters, since several of our authors, especially Cantor and Peirce, argue that density is a necessary condition of any continuous system.

Despite its seeming inapplicability to the current debate, we will see echoes of Aristotle's argument throughout our story. Whether a line can be composed of points, or whether any continuum can be composed of indivisibles, is a fundamental debate in many discussions of continua. Aristotle's argument was repeated and elaborated on throughout the medieval period, and in Chapter 5, we will see that similar considerations led du Bois-Reymond to a conclusion that is very different from Aristotle's. In Chapter 6, we will see Peirce used a modified version of Aristotle's tripartite distinction to define his own system of mathematical continuity, even though he later came to agree with Aristotle that continua cannot be composed of indivisibles.

Before leaving Aristotle, it is interesting to note that after forwarding his tripartite argument against a compositional continuum, he uses similar reasoning to prove what he has already stated, that a continuum cannot be decomposed into indivisibles:

> Moreover, it is plain that everything continuous is divisible into divisibles that are always divisible: for if it were divisible into indivisibles, we should have an indivisible in contact with an

indivisible, since the extremities of things that are continuous with one another are one and are in contact.[9]

Thus, imagine a division in a continuum that produced, instead of two parts that were themselves continuous and infinitely divisible, two parts that were indivisible. These indivisibles would be connected continuously (as they are the product of a division on a continuum), but this implies that they share borders. As Aristotle demonstrated in the last argument, indivisible entities have no borders to share, thus such a division is impossible, and every division on a continuum results in things which are themselves divisible. As we shall see in the next section, Archimedes made significant mathematical advances by rejecting just this characterization of continuity, and instead viewing continua as being divisible into indivisible entities.

2.3. Archimedes and the Infinitely Small

Archimedes of Syracuse (281–212 BCE) assumed that continuous things were in fact composed of indivisible parts; and more importantly, that a continuum could be divided into indivisibles. A plane, for example, could be viewed as composed of infinitely many lines. While lines are certainly divisible along their length, they are not divisible along their height, as they lack that dimension; in this sense, they are indivisible elements. The assumption of an entity containing infinitely many indivisible objects was key to Archimedes' "method of exhaustion," a startlingly accurate system he used to measure seemingly immeasurable geometric figures and shapes.[10] The method of exhaustion is similar in many respects to the calculus of Newton and Leibniz, and in fact, this method heavily influenced the mathematicians of the sixteenth and seventeenth centuries in their development of systems of calculation which would set the groundwork for the development of calculus itself.

The method of exhaustion involved first inscribing the figure to be measured with infinitely many figures of a specified shape. For example, to measure a parabolic segment, he first inscribed a triangle within it.

---

[9]Ibid., 231b16–19.
[10]For a thorough description of the method of exhaustion, see Boyer [1959, p. 50ff.].

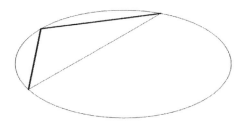

FIGURE 2.1. Method of Exhaustion, 1ˢᵗ Inscription

Next, he would inscribe smaller triangles, which had for their bases the sides of the original triangle.

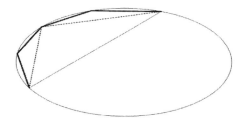

FIGURE 2.2. Method of Exhaustion, 2ⁿᵈ Inscription

The sides of these triangles were then used as the bases for new inscribed triangles, and in this manner, Archimedes obtained "a series of polygons with an ever-greater number of sides. ... He then demonstrated that the area of the nth such polygon was given by the series

$$A\left(1 + \frac{1}{4} + \frac{1}{4^2} + \cdots + \frac{1}{4^{n-1}} + \frac{1}{3}\frac{1}{4^{n-1}}\right) = \frac{4}{3}A$$

where A is the area of the inscribed triangle having the same vertex and base as the segment,"[11] the segment being the original parabolic segment which was to be measured.

---

[11]Ibid., p. 52.

Thus, we view the area inside the parabolic segment as composed of infinitely many infinitely small triangles, creating a polygon of infinitely many sides. Similarly, Archimedes would circumscribe a triangle around the outside of the polygon, and circumscribe infinitely decreasing triangles between this larger triangle and the external walls of the parabolic segment, creating another polygon of infinitely many sides. One could use the equation in the Boyer quotation above with the theory of limits to find the length of the curve and the area of the parabolic segment, but Archimedes did not view this simultaneous inscribing and circumscribing of polygons with smaller and smaller sides as approaching a limit. Rather, after inscribing and circumscribing his two different infinitely-many sided polygons, he argued by a double *reductio ad absurdum* that the parabolic segment above "could be neither greater nor less than $4/3A$."[12]

Archimedes did not operate with a concept of a limit, nor did he attempt to formulate any generalized means of finding the area of such figures, but rather restricted himself to inscription and circumscription, using different inscribed figures as each shape demanded. Thus, although he was highly influential on the mathematicians who helped give rise to the calculus, he himself can not be viewed as having invented the calculus. It is interesting, however, to note that a sort of infinitesimal calculation was in use as early as Archimedes. In fact, had medieval philosophers and mathematicians been less influenced by Aristotle to the exclusion of Archimedes, it is possible that calculus or something like it may have been invented much earlier, building on these Archimedean concepts.

2.4. The Medieval Debate on Motion, Change, and Continuity

Even before medieval scholars had access to Aristotle's *Physics* and its analysis of motion as implying continuity, they were highly interested in the peculiar logical qualities of the words "stop" and "start;" i.e., they saw philosophical problems precisely where motion changes into non-motion, and the reverse.[13] An analysis of this medieval debate will help set the ground for the philosophical problems that arose with the introduction of calculus itself.

---

[12]Ibid., p. 52.

[13]Norman Kretzmann supports the thesis that beginning and ceasing were important to the medievals before Aristotle's *Physics* was widely available throughout Western Europe. See Kretzmann [1976].

To understand the difficulties inherent in the concepts of starting and stopping, consider the sentence, "This ball begins to roll." It is perfectly straightforward, on the face of it; a simple present-tense positive statement. Yet embedded within the sentence is an implicit negation, as well as an implicit reference to either the past or the future (or both), depending on how the sentence is analyzed. For "This ball begins to roll" implies that an instant ago, the ball was *not* rolling (the implied negation), and also implies something about both the past (that the ball *was* not rolling), and the future (that the ball *will*, at least in the immediate future, continue to roll, as motion must take place in time). "The ball ceases to roll" has parallel complexities; it implies that an instant ago, it was rolling, and an instant from now, it will no longer do so, and thus we have an embedded negation and references to the past and the future, all contained in a seemingly simple present-tense positive statement.

Thus, early medieval discussions dealt with the complex logical implications of the notions of starting and stopping. Once the medieval philosophers became familiar with Aristotle's *Physics*, the somewhat complex logical puzzle of starting and stopping exploded into full-blown metaphysical conundra. For all beginnings and endings imply change; and change, as analyzed by Aristotle and medieval philosophers, is a type of motion. For Aristotle, and thus for later medieval philosophers, motion implied continuity.

Medieval philosophers inherited Aristotle's analysis of continuity almost wholesale when they gained access to the *Physics*. Thus, they inherited the ideas that space, time, and motion were all continuous, and that continua were infinitely divisible. The first idea meant that every discussion of natural philosophy had to include some discussion of continuity, as natural philosophy almost always dealt with change of some sort. Thus, for example, when William of Ockham (1287–1347) tackles the question "Do angels move?" he must not only reckon with the Bible, with the metaphysical status of angels, and with the meaning of motion itself, he must also grapple with the continuity of motion, time, and space.[14] The second idea meant that the nature of infinite divisibility needed to be understood thoroughly before natural philosophy could be understood fully.

---

[14]See Ockham [1991, Question 5]. This puzzle is described in footnote 12 of Chapter 1.

To see how the concept of continuity influenced the medieval debate on motion, let us return to the example of a rolling ball. Medieval scholars wished to know what occurs during the very instant the ball begins to roll. For consider: the ball begins to roll at time t. Thanks to the law of excluded middle, we know that the ball either is or is not rolling at t. Furthermore, the time span when the ball is not rolling must be adjacent to the time span when the ball is rolling (for if there were any time in between states, the ball would neither be rolling nor not rolling at that time); and the two periods cannot overlap (for if they did, then at some period it would be both rolling and not rolling – a contradiction).

Yet given the assumptions that (i) time is a continuum, (ii) a continuum cannot be composed of indivisibles, and (iii) any indivisibles which appear in a continuum cannot be next to each other,[15] we seem to have reached a difficulty. For if the ball begins to roll at time t, what happens immediately before time t with respect to the ball? There cannot be both a first instant of rolling and a last instant of non-rolling, for that implies either that instants are next to each other, which Aristotle argued can never happen, or that there is a period of time when the ball is neither rolling nor non-rolling, which is a contradiction.[16]

The difficulty was approached in various ways by various philosophers. Though it was not philosophically fashionable to reject Aristotle's key views, a small group did so, and argued that time and space were in fact composed of indivisibles. Walter Chatton (c. 1290–1343), for example, argued that a line was not only composed of points, but was composed of finitely many points. Other philosophers, such as Walter Burley (c. 1275–1344/5), analyzed change by using something similar to limits, thus claiming that the rolling ball can have a first instant of movement but not a last instant of non-movement. It is interesting to see the concepts of continuity and limits philosophically connected centuries before limits were formally introduced in mathematics. Ockham, on the other hand, argued strongly against the existence of indivisibles that his contemporaries supported, drawing not only on Aristotle's arguments against indivisibles, but on mathematical and geometrical arguments first proposed by

---

[15]Recall from Section 2.2, Aristotle's statement that "two points must be next to each other" was not a statement Aristotle affirmed, but was supposed to follow from the *reductio* assumption that the straight line is composed of indivisible points. This statement was rejected in the course of the argument.

[16]This discussion of the problem is influenced by the discussion in Paul Spade's "How to Start and Stop: Walter Burley on the Instant of Transition," [Spade, 1994, p. 193–221].

2.4. THE MEDIEVAL DEBATE ON MOTION, CHANGE, AND CONTINUITY 27

al-Ghazali (1058–1111) and popularized in the West by John Duns Scotus (c. 1265–1308).[17]

An issue related to stopping and starting is the question of whether or not motion is possible at an instant. Aristotle himself discussed this question, and concluded in the negative, but his discussion focused on whether a motion can take place *entirely* within an instant. A more complicated issue is whether motion can happen *at all* in an instant. Think, for example, of our rolling ball; we know it is indeed in motion while it rolls, but is it therefore in motion in every particular instant throughout the duration of its travel? It seems as though it should be the case that, if an object has a property throughout a period of time, it should also have that property through any part of that period, even a part as short as an instant. Yet some accounts of motion prohibit such an understanding. Ockham, for example, defined the motion of x as meaning that x is in some place at time t, and at a different place at a later time, t_1. Some codicils apply to ensure that the object x did not somehow, for example, cease to exist during the time period, but the notable feature of Ockham's definition is that it is only possible to say whether something has moved after a span of time has elapsed. An instant is not a span of time, thus to say something is in motion at an instant is therefore nonsense, even if that instant occurs during the time period between t and t_1, that is, during the time at which the object was in motion.

Of course, the idea that an object cannot have motion at an instant is not only counter-intuitive (what does a moving object do at such an instant? Stop?) but it is also antithetical to the basic principles of modern analysis, where we not only claim that an object in motion is also in motion at an instant, but we are able to calculate the object's velocity at that instant. Fortunately, not every medieval philosopher agreed with Ockham. Chatton, for example, (who disagreed with much of what Ockham wrote) presented an analysis of motion which left open the possibility of instantaneous motion.

Thus, continuity was very much a live and lively issue for philosophers during the medieval period. The assumption of the continuity of space, time, and magnitude influenced discussions about motion, starting, and stopping, and also

---

[17]See Ockham's first quodlibet, [Ockham, 1991, Question 9], "Is a line composed of points?"

appeared in geometrical discussions, such as whether or not a line is composed of points. This period is interesting not only in its own right, but because these discussions set the stage for early modern mathematical developments, from Cartesian geometry to the development of the calculus.[18]

2.5. Analysis

It is notable that up until this point in our history discussion, continuity was not in the main a mathematical property. It was mostly a physical and metaphysical issue; geometry certainly touched on such topics, but arithmetic never did.[19] This was to change, of course, with the introduction of real number theory, Cartesian graphs, and similar mathematical advances; but no event or theory changed the face of mathematics quite so much as the introduction of the calculus.

A fascinating feature of the century leading up to the invention of the calculus was a change in attitude towards Aristotle. Elie Wiesel (b. 1928) has said many times of his own religious faith, "With God or against God, but never without God."[20] Medieval scholars had a similar approach to Aristotle, analyzing his works, arguing with him, but almost always mentioning him in some form or fashion. This changed in the sixteenth century. While Aristotle was still studied in this century, a new and growing group rejected both Aristotle and the Aristotelianism of the medieval age.[21] Mathematicians of this age rejected Aristotle's argument that mathematics was the science of counting and that number began with 1. The door was thus opened for the inclusion of zero, borrowed from Arab mathematicians, negative numbers, fractions, irrational

---

[18]The discussions of Burley, Ockham, Chatton and others in the 14th century and earlier influenced a group of logically and mathematically minded gentlemen often referred to as the "Oxford Calculators" or the "Mertonian Calculators", who worked on similar issues of space, time, and motion in the late fifteenth and early sixteenth centuries; the Calculators in their turn influenced the science of Leonardo da Vinci and Galileo. These medieval ponderings were thus not inconsequential to the history of mathematics and science. See Edith Dudley Sylla's article "The Oxford calculators," [Sylla, 1982].

[19]The exception to this, of course, is Archimedes, who used his concept of a compositional continuum to calculate the area of irregular geometrical shapes.

[20]As quoted in an interview in the *U.S. News and World Report*, October 27, 1986, p. 68.

[21]See Boyer [1959, p. 96].

2.5. ANALYSIS

numbers, and imaginary numbers, all of which were newly accepted as types of numbers by Western mathematicians of the period.[22] This partial rejection of Aristotle also opened the door for reappraisal of Archimedes, and it was Archimedes who burst into prominence in certain intellectual circles, his works appearing in numerous editions, his method of exhaustion thoroughly studied, repeated, and expanded upon.

Mathematicians in this period who did notable work on furthering the method of exhaustion were Simon Steven (1548–1620), Luca Valerio (1552–1618), and Johannes Kepler, (1571–1630). In each case, limit theory was approached, but never reached. Kepler in particular innovated considerably on Archimedes' method. In computing the volume of various solids, Kepler assumed them to be composed of infinitesimal shapes of various kinds, not always assuming these infinitesimals to be indivisibles.[23] Bonaventura Cavalieri (1598–1647) took Kepler's methods even further (though he denied any inspiration from Kepler's work), writing the highly influential *Geometria indivisibilibus* in 1635.

Thus, by the mid-seventeenth century, the time was ripe for analysis; so ripe that two apples fell almost simultaneously from the calculus tree. Newton and Leibniz, apparently independently, introduced a method of mathematics which went by several names – infinitesimal analysis, differential calculus, etc. – and which today is referred to in college courses simply as "the calculus," as though there was never any other. Among other things, this new calculus allowed mathematicians to calculate the tangent to any curve with remarkable precision. Applied to science, the theory allowed the measurement of trajectory at a point, and of motion at an instant. The continuity of time and space became a feature of mathematics at large, rather than being a feature only of geometry, as continuous functions were mapped, plotted, and calculated; and mathematics suddenly received all the metaphysical concerns of continuity, along with a whole host of other philosophical concerns.

Chief among the metaphysical concerns associated with calculus was "What, precisely, are we measuring?" To demonstrate the problem explicitly, a concrete example can help. The advances Descartes made in algebra allowed

---

[22]Ibid., p. 97–98.
[23]Ibid., p. 109.

curves to be calculated with precision, and algebraic equations to be plotted as curves. However, calculating the slope of a line was still only possible under certain circumstances. Imagine, for example, that we have a curve on a graph, the equation of the curve, and a particular point on the curve – call the point P_1, with coordinates (a,b). We wish to find the slope of the tangent line that intersects the curve at P_1. If we had another point on the tangent line, we could easily calculate its slope: if we had point (a,b) and also (c,d), the slope m is found with the equation $m = {d-b}/{c-a}$. However, with only one point, this formula is useless.

One of the fundamental advances in calculus is that it allows us to find this slope, given only the point P_1 and the equation of the curve, in roughly the following manner. Find a second point on the curve, relatively close to P_1 – call it P_2 – and draw a straight line through these two points. We can use the above formula to figure the slope of this line, but this is not the tangent line, it is rather a secant – it crosses our curve at two points rather than intersecting it at only one. Now, we find a point on the curve closer to P_1 – call it P_3 – and draw another secant line. The slope of this line is closer to the slope of the tangent line than the previous secant, since P_3 is nearer to P_1, but it is still not the tangent itself. We can keep choosing points closer and closer to P_1, and with every step closer, the slope of the resulting secant line becomes a more accurate estimation of the slope of the tangent.

However, notice what happens if we choose the point P_1 itself, rather than a point near it – that is, notice what happens if the distance between the points on the curve collapses. If we plug the coordinates of P_1 into the slope formula we are left with the equation
$$m = \frac{b-b}{a-a}.$$
Not only is this not informative, it is nonsensical; the equation requires that we divide by zero, which we cannot do. Thus, ideally, the point we choose to use as an estimation for the tangent line itself must be *as close as possible* to P_1 without actually being *equal* to P_1. The closer the point is, the better the estimation, but some distance must be maintained to avoid division by zero. In fact, if we could find a point P_n that was *infinitely* close to P_1, the secant line would no longer give us an estimation of the tangent line – rather, we could calculate the actual slope of the tangent. Both Newton and Leibniz were sure this infinite closeness could be suitably formulated and used in calculations.

2.5. ANALYSIS

Their original attempts to do so were met with philosophical and mathematical skepticism, but notably, though the two men used very different means of formulating this infinite closeness, their calculations produced the same results.

However, characterizing with exactitude this part of the calculation – the place where P_1 and P_n become infinitely close without becoming equivalent – proved a difficult task, and one which created much ado among the mathematical community. What, precisely, was this "vanishing point"? Of what did it consist? Was it indeed an infinitely small ratio? Did the idea of an infinitely small ratio make any sense? Did it make any more sense than dividing by zero? Were they, after all, no more than "ghosts of departed quantities" as Berkeley charged? Infinitesimals had never been universally accepted among the philosophical community, and infinitesimal quantities were no less controversial.

The two founders of the calculus dealt with these open metaphysical and mathematical questions in different ways. Newton was, at bottom, a scientist in need of a tool, and the tool was needed to analyze motion. He saw continua as created by motion, and saw the difference between these two points (P_1 and the very close point used to calculate the tangent) as an infinitely small moment in time.[24] Thus, imagine the curve as generated by the motion of some entity, and imagine a very accurate stopwatch. We start the watch when the entity reaches P_1, and stop it a an infinitely small moment later, when it reaches P_n. The distance traveled in that moment is, roughly speaking, a fluxion. However, Newton ultimately became dissatisfied with this formulation, and especially with the explicit reference to infinitesimal quantities; he eventually rejected it in favor of a theory of ratios, which kept the references to motion, but which more closely resembled modern limit theory.[25]

Leibniz thought of the difference between P_1 and P_n as an infinitely small distance,[26] but it bothered him that he did not have a clear understanding of what an infinitely small distance *was*. After much thought, he concluded that infinitesimals were not actual numbers, but idealized fictions, useful for

---

[24]See John Bell's "Continuity and Infinitesimals," [Bell, 2005, Section 4].

[25]Ibid, Section 4.

[26]Leibniz does use the word "infinitesimal" (in Latin, *infinitesima*) as a noun, but more frequently uses the term "infinitely small" (*infinite parvum*) as a substantive or an adjective.

mathematics.[27] An 'infinitesimal' was no more than a way of speaking about the distance between P_1 and P_n, a way of referring to it and moving on with the calculation. However, Leibniz's subsequent attempts to define this useful fiction rigorously were problematic. He presented an elegant set of rules governing his calculus, and insisted that infinitesimal quantities were to "obey precisely the same algebraic rules as finite quantities."[28] This insistence led directly to the conclusion that infinitesimal quantities must be treated "in the presence of finite quantities, *as if* they were zeros."[29] It is clear from our above example that if an infinitesimal quantity is treated as if it were zero, then the introduction of an infinitesimal difference between P_1 and P_n no longer accomplishes what it was meant to – that is, no longer keeps a large enough distance between the two points to avoid division by zero.

Perhaps because it was relatively well defined, Leibniz's calculus dominated the continent for several decades, gaining faithful followers such as Guillaume de l'Hôpital (1661–1704), who developed it further. Yet, the mathematical and philosophical unease about infinitesimals led mathematicians to look for a new foundation. The proto-limits mentioned in Newton's calculus held some appeal as an alternative, and more precise definitions of them were sought. Great strides toward ridding the calculus of vague definitions and useful fictions were made by Augustin Cauchy (1789–1857), who developed a rigorously defined theory of limits. As Boyer wrote,

> In the work of Cauchy, however, the limit concept became [...] clearly and definitely arithmetical rather than geometrical.[30]

To return briefly to our tangent example above, Cauchy's limit theory allowed us to calculate the tangent at P_1 arithmetically, without reference to geometry at all. Rather than positing a second number P_n existing an infinitesimal distance away from P_1, we could instead calculate a variable quantity P_x, which infinitely

---

[27]Interestingly, the attempt to define infinitesimal quantities in mathematics ultimately led Leibniz to reject their mathematical existence, but consideration of the same issues led him to postulate the actual existence of infinitely small entities called "monads," on which his metaphysics is based. See Bell [2005, Section 4]. For Leibniz's proof that infinitely small things are fictions, see Leibniz [2001, Aiii52].

[28]See Bell [2005, Section 4].

[29]Ibid., Section 4.

[30]See Boyer [1959, p. 272–273].

approached P_1 but never reached it. In other words, limit theory allowed the calculation of the tangent without the necessity of positing an infinitely small magnitude.

Limit theory was further developed and refined, and calculus gained wider acceptance. However, limit theory did not have universal approval among mathematicians and philosophers. Some were suspicious of limits, and many were still convinced that infinitesimals had a place in calculus and in mathematics in general. Paul du Bois-Reymond, Charles Sanders Peirce, Otto Stolz (1842–1905), and Charles Dodgson (1832–1898) are some of the mathematicians who attempted to develop infinitesimal theory further, even after limits had a widespread following.

2.6. Conclusions

Limit theory proved to be a mathematically powerful system, one which has been used to great effect since its adoption: one need only think of the advances in physics, astronomy, chemistry, and virtually every scientific field which uses calculus to be convinced of this fact. Philosophically, limit theory solved the troublesome need to refer to infinitesimals (though as was briefly noted above, and as shall be seen in more detail below, not everyone believed infinitesimals to be troublesome or unnecessary), but necessitated the extension of the concept of continuity into the realm of systems of numbers. For, after Cauchy, calculus was freed from its necessary connection to geometry, and one could calculate motion at an instant without reference to curves and lines, but only if number systems themselves contained the property of continuity. If P_x (now a variable quantity, not a point) is to approach some number P_1 infinitely and continuously, it can only do so if the numbers themselves form a continuous set.

The history of various theories of continuity is interesting in its own rights, but it also is important for the chapters which follow, as many of the difficulties inherent to mathematical continuity in particular stem directly from the history behind its invention. Of primary importance is the debate between those who believe a continuum can never be composed of indivisible elements, and those who think it can. At the dawn of this debate, Aristotle argued that continua can never be composed of indivisibles, while Archimedes simply treated continua as though they were composed of indivisibles, and made mathematical advances based on that assumption. In the medieval period, Aristotle was favored in many things, but indivisibilists still argued their case vigorously. In the early-modern period, it was a rejection of Aristotle that led to the expansion of the very concept of number, but he still had his influence: Leibniz himself was convinced that Aristotle's argument was correct, and that continua could not be composed of indivisibles.[31]

This ongoing debate is relevant to the subsequent chapters as it is one of the main issues all four figures will address in one way or another. By discussing whether number systems can be continuous, the debate itself changes, as it is not immediately clear whether a number is an indivisible, or what it would even mean to refer to a number as indivisible or not. However, those who argue in favor of a mathematical continuum are compositionalists in the sense of believing that continua can be composed of atomic elements. The nature of those elements, as well as the manner of composition, become important issues in determining whether a continuum can or cannot be composed of these elements. Dedekind and Cantor believed that continua could indeed be composed of atomic elements; both believed that the real numbers form a continuum without the necessity of adding any other mathematical or non-mathematical element.

We shall also see, in the chapters that follow, what became of the debate about the nature and mathematical formulation of infinitesimals after the introduction of limits. Du Bois-Reymond believed that limit theory was only possible if the continuum over which limit theory ranges contains both numbers (or points) and infinitesimal intervals. Peirce believed that both mathematical and non-mathematical continuity requires infinitesimals, though these different continua use infinitesimals in different roles. Peirce also believed that infinitesimals were useful mathematical entities in their own right, not only adding continuity to otherwise discontinuous systems, but allowing for certain calculations and fine-grained distinctions impossible with limit theory alone.

Thus, as our four mathematicians are considered one by one, the reader should keep in mind the following historical lessons. First, the definition of

---

[31]See Bell [2005, Section 4].

2.6. CONCLUSIONS

continuity itself, as presented by Aristotle and medieval philosophers, must change as it applies to systems of numbers; however, it must remain close enough to be identifiable as continuity. Second, defining continua as composed of atomic elements has always had its detractors, but has had many supporters who have used the composition to great effect. And finally, the role of infinitesimal elements in continua has been an issue since before infinitesimals were mathematically introduced, and their mathematical introduction allowed the birth of calculus, but almost single-handedly destroyed it in its early years.

Though each of our authors approached the concept of continuity differently, each addressed these three issues: the definition of continuity (sometimes distinguishing the definition of continuity in general from the definition of mathematical continuity in particular), whether continuity could be composed of atomic elements, and the role of infinitesimals in continuity.

CHAPTER 3

Richard Dedekind

3.1. Biography and Introduction

Julius Wilhelm Richard Dedekind was born October 6, 1831, in Braunschweig, in what is now Germany. Early in his career he was interested in science, particularly physics and chemistry. He soon became disenchanted with the imprecise nature of physical science, and his interest in mathematics increased. At the age of 16, he entered the Collegium Carolinum, and two years later he attended the University of Göttingen, where he studied physics with Wilhelm Weber (1804–1891) and mathematics with Carl Friedrich Gauss (1777–1855); in fact, he was Gauss's last student. He earned his doctorate in 1852, and spent the next two years learning the latest mathematical developments.

In 1855, Dedekind began to take courses in number theory, definite integrals, and partial differential equations from Lejeune Dirichlet (1805–1859). He associated with Dirichlet almost daily, writing of this association that he was "for the first time beginning to learn properly."[1] Around this time he studied the works of Evarist Galois (1811–1832), and was among the first to lecture on Galois Theory. In 1858, Dedekind was appointed to the Polytechnikum in Zürich, where he first taught differential and integral calculus. In 1862, Dedekind moved to the Polytechnikum in Braunschweig, and remained there until his retirement in 1894. He died in 1916, in the same city in which he was born.

When he began to teach calculus in Zürich, he was troubled by the necessary references to geometry that were standard in teaching methods. Recall

---

[1]Written in an 1856 letter. See Dedekind [1981].

from Chapter 2 that Cauchy had arithmetized the theory of limits; still, the introduction of these limits depended on references to geometrical curves and tangents. Pedagogically, of course, geometric figures and diagrams were most helpful, but Dedekind was worried that calculus might be thought to depend essentially on these diagrams. At times it seemed to him as though the whole of analysis itself might rest on such an unscientific and non-rigorous base as geometric intuitions. Dedekind found this unacceptable, and he became determined to fix it; "I made the fixed resolve to keep meditating on the question till I should find a purely arithmetic and perfectly rigorous foundation for the principles of infinitesimal analysis."[2] He eventually achieved this end to his own satisfaction, and published his result in his 1872 essay, *Stetigkeit und irrationale Zahlen*.

The elements of calculus which most troubled him were those which involved the assumed but not yet proven continuity of the real numbers, and the geometrical and infinitesimal references often used in proving the theorem that "every variable magnitude which approaches a limiting value finally changes by less than any given positive magnitude."[3] In fact, this theorem appears at the end of *Stetigkeit und irrationale Zahlen*, and is proved without reference to infinitesimals or geometrical intuitions, relying instead upon his newly established principle of continuity.

Dedekind is an important figure in several regards. Dedekind's method of defining real numbers is closely related to other algebraic methods, but stands out as remarkably self-contained and straightforward. Dedekind's real numbers are also notable because he explicitly linked them to his principle of continuity, defining continuity itself in terms of them. This chapter will first explain Dedekind's real numbers and his principle of continuity in Section 3.2. Section 3.3 will investigate Dedekind's conception of number itself, and Section 3.4 will contain a fuller discussion of continuity and infinitesimals in relation to his real number theory.

---

[2] See Dedekind [1963]. This small volume contains translations of both *Stetigkeit und irrationale Zahlen* and *Was sind und was sollen die Zahlen?*

[3] See Dedekind [1963, p. 26].

3.2. Dedekind Cuts and the Principle of Continuity

Dedekind created real numbers through cuts on the collection of rational numbers. In this section, I shall first describe Dedekind's method of creating cuts, and therefore irrational numbers. I shall next outline his proof that the collection of real numbers thus produced satisfies his intuitions of continuity.

A cut is a complete division of a collection of numbers, such that every member of the first half is less than every member of the second half. The cuts which define irrational numbers are thus divisions of the collection of all rational numbers. The rationals can be totally ordered: for any two rationals a and b, either $a \leq b$ or $a \geq b$. The rationals are also dense; between any two rationals there is a third. We can thus divide the rationals into two classes,[4] A and B, such that every member of A is less than every member of B. In fact, every rational number produces just such a division: for any rational number n, every other rational number is either greater or less than n; thus, by arbitrarily placing n into the lower class, we can form

$$A = \{x \mid x \leq n\}$$

and

$$B = \{x \mid x > n\}.$$

Every member of A under this assignment is clearly less than every member of B, and thus, this division counts as a cut.

For any such cut (P, Q), if the set P has a greatest rational number, then we say this number *produces* the cut. Thus, in our example above, the rational number n – the greatest number of A – produces the cut (A, B). Note that we could just have easily put n into the second class rather than the first, forming cut (A', B') such that

$$A' = \{x \mid x < n\}$$

and

$$B' = \{x \mid x \geq n\}.$$

---

[4]Modern presentation of Dedekind cuts often refer to two "sets." Dedekind himself had a theory of systems, which is similar to the modern conception of sets, but differs from it in important respects. Thus, I will use either "system" or the more neutral terms "class" or "collection" to avoid confusion.

In this case, n still produces the cut (A', B'), since it is also true that for any cut (P, Q), if the class Q has a least number, then that least number produces the cut. Notice, though, that as there are infinitely many numbers between any two rationals, and as every rational must belong to either P or Q, it is not possible for P to have a greatest member and Q to have a least member in the same cut.

However, it is possible for neither P nor Q to have a greatest or least member, respectively, and thus, possible for a cut on the rationals to be produced by no rational number.[5] As Dedekind demonstrates, there are infinitely many cuts of this latter sort, infinitely many divisions of the rationals into cuts (P, Q) where every member of P is less than every member of Q, yet P does not have a greatest, nor does Q have a least, rational number.[6] Thus:

> Whenever then, we have to do with a cut (A_1, A_2) produced by no rational number, we create a new, an *irrational* number a, which we regard as completely defined by this cut (A_1, A_2); we shall say that the number a corresponds to this cut, or that it produces this cut.[7]

Thus, the irrationals are born. Essentially, wherever a gap appears in the rationals, we fill it with an irrational; this irrational is simply and completely determined by a cut. The system of real numbers is thus the collection of all cuts on the rational numbers, both those produced by rationals and those (irrationals) which are not.[8]

---

[5]Though it is question-begging, as the purpose here is to construct the irrational numbers, the easiest way to imagine such a cut is to consider the division of the class of rational numbers precisely at the point where the square root of two would fall. Every member of the lower class would be less than every member of the upper class, and yet, as the square root of two is not a rational number and thus not part of either class of rationals, the lower class would not have an upper bound, nor would the upper class have a lower bound.

[6]The mathematical proof of this occurs in Dedekind [1963, p. 13–35].

[7]Ibid., p. 15.

[8]Modern presentations of Dedekind cuts equate the irrational number with the two sets of rationals themselves, or, more frequently, simply the lower of the two sets. See, for example, planetmath.org/DedekindCuts.html. However, notice that Dedekind leaves open the question of whether the pair of systems A_1 and A_2 are themselves equivalent to the irrational number, or whether an irrational number independent of these sets is created which thus can be said to produce the cut.

3.2. DEDEKIND CUTS AND THE PRINCIPLE OF CONTINUITY

Dedekind went on to argue that these irrationals have at least some of the properties we would like them to have, and that we can calculate with these irrationals in just the ways we need to. The basic method behind these proofs is simple:

> To reduce any operation with two real numbers α, β to operations with rational numbers, it is only necessary from the cuts (A_1, A_2), (B_1, B_2) produced by the numbers α and β in the system R to define the cut (C_1, C_2) which is to correspond to the result of the operation, γ.[9]

In just this way, he proves that cuts work as we would expect them to in terms of addition, and indicates how similar proofs could establish the ability of real numbers (as defined by cuts) to perform "the other operations of the so-called elementary arithmetic [...] differences, products, quotients, powers, roots, logarithms," and eventually, the proofs of theorems.[10] In other words, these irrationals, determined by cuts in the rationals, have many of the properties we desire.

Dedekind argued that the collection of real numbers – the rationals, combined with the irrationals constructed using cuts – form a continuous set. Dedekind offered his definition of continuity early in the essay, when considering the geometry he was trying to move away from. First, in the section entitled "Continuity of the Straight Line," Dedekind wrote that "in the straight line L there are infinitely many points which correspond to no rational number,"[11] and thus that "the straight line L is infinitely richer in point-individuals than the domain R of rational numbers in number-individuals."[12] While continuity is not merely a matter of size, the bigger somehow being the more continuous, clearly this difference shows that something about the rationals is lacking when we attempt to measure the line.[13] Thus, whether rationals are 'continuous' or not (Dedekind would say "not," of course), they are definitely insufficient if, as

---

[9]Ibid., p. 21.
[10]Ibid., p. 22.
[11]Ibid., p. 8.
[12]Ibid., p. 9.
[13]The reader may recall the example in Section 1.4, which demonstrated that a comparison between the geometrical straight line and the rational number line could find gaps in the latter – places where there simply is no number.

Dedekind says, "we try to follow up arithmetically all phenomena in the straight line."[14] The rationals do not allow for complete arithmetic analysis of the line, but we need a number system as least as complete, as least as continuous, as the line.

And indeed, Dedekind finds the very essence of continuity in the straight line:

> If all points of the straight line fall into two classes such that every point of the first class lies to the left of every point of the second class, then there exists one and only one point which produces this division of all points into two classes, this severing of the straight line into two portions.[15]

Thus, claimed Dedekind, if you cut a geometrical straight line, you must necessarily cut it *at a point*, and at only one point. It is impossible to cut a geometrical line between points, or to have two points determine the same division on the line.[16] This, not infinite divisibility or metaphysical smoothness, is what Dedekind took to be the feature that distinguishes continuous from non-continuous things, or at least it is the feature that distinguishes continuous from non-continuous geometrical objects.

The resemblance between this "essence of continuity" and Dedekind-defined cuts is obvious. Replacing a few words, we could say that continuity of the reals happens on the following condition: if all real numbers fall into two classes such that every number of the first class is less than every number of the second class, then there exists one and only one number which produces this division of all numbers into two classes, this severing of the reals into two portions. Dedekind cuts establish half of this condition: the creation of irrational numbers through Dedekind cuts ensures that every such division is produced by a number (if there is no rational number at that cut, we create an irrational one). If in addition it could be proven that each cut is produced by only one number, this condition would be fulfilled, and the real numbers would exhibit this essence of continuity.

---

[14]Ibid., p. 9.
[15]Ibid., p. 11.
[16]As we shall see in Chapter 5, Paul du Bois-Reymond thought quite the opposite.

3.2. DEDEKIND CUTS AND THE PRINCIPLE OF CONTINUITY

Dedekind argues that exactly one number produces each cut on the reals as follows.[17] First, consider a cut (Φ, Ψ) on the reals – that is, consider a division of the real numbers such that every real number belongs either to Φ or to Ψ, and that every member of Φ is less than every member of Ψ. At the same time we produce cut (Φ, Ψ), we also produce cut (P, Q) on the rationals, P containing all the rationals in class Φ and Q containing all the rationals in the class Ψ. As we saw above, every cut on the rationals is produced by a number; if there is no rational which produces it, we create an irrational and say that this irrational produces the cut. Thus, let a be the "perfectly definite"[18] real number which produces cut (P, Q). It will become clear that a also produces cut (Φ, Ψ). Now, for the purposes of *reductio*, we attempt to find a second real number, distinct from a, which also produces the cut (Φ, Ψ). For any $b \neq a$, either $b < a$ or $b > a$. Since the rationals are dense, there are infinitely many rationals c between b and a, and thus, if $b < a$, it is also the case that $c < a$. Since a is the number that produces cut (P, Q), all numbers less than a belong to P, and therefore c must belong to P; since all members of P are also members of Φ, c also belongs to Φ. Since c belongs to Φ, and $b < c$, b also belongs to Φ. If, on the other hand, $b > a$, then $c > a$, and c is therefore a member of Q, and thus of Ψ. Thus, every number b different from a must either belong to Φ or Ψ. Furthermore, b cannot be the greatest number of Φ, for if b is a member of Φ, then there are infinitely many numbers greater than b which are also members of Φ. Parallel reasoning shows why b cannot be the least member of Ψ, and thus, b cannot produce the cut (Φ, Ψ). Therefore, only one number can make the cut; only one number produces the separation between Φ and Ψ.

Thus, Dedekind proved that for every cut, there is one and only one real number which produces the cut. These cuts correspond to Dedekind's essence of continuity, and so, according to Dedekind, the real numbers thus produced form a continuum. The philosophical question before us is not whether Dedekind's cuts produce a real number continuity according to his own definition; they most certainly do. The question is whether this definition captures our intuitions of continuity itself. Given that *continuous* is a word with many aspects, a word seemingly applicable in physics, metaphysics, algebra and geometry, it is reasonable to expect it to take on different shades, or even different meanings,

---

[17]See Dedekind [1963, p. 20–21].
[18]Ibid., p. 20.

in different fields. It is natural for us to need continuity to function differently in different places, and for its definitions to change accordingly; not every use of the word will fall directly out of the dictionary definition. Jules Henri Poincaré, for one, thought that mathematical continuity was very different from our "ordinary conception" of continuity:

> The continuum thus conceived is nothing but a collection of individuals arranged in a certain order, infinite in number, it is true, but external to each other. This is not the ordinary conception, in which there is supposed to be, between the elements of the continuum, a sort of intimate bond which makes a whole of them, in which the point is not prior to the line, but the line to the point. Of the famous formula, the continuum is unity in multiplicity, the multiplicity alone subsists, the unity has disappeared.[19]

In other words, Poincaré is objecting to the compositional nature of mathematical continua. As we shall see in later chapters, Cantor, du Bois-Reymond, and Peirce attempt to formulate mathematical continua which characterize the "intimate bond which makes a whole of" the elements of the continuum, but Dedekind himself has no such bond. He simply characterizes continuity in terms of completeness: there are enough real numbers such that anywhere you wish to divide them, you will divide at a number, there are no gaps. Yet these numbers have no intimate or non-intimate connection between them; they simply exist in this collection, in this order, in this completeness, and that is enough, for Dedekind, to call the collection continuous.

The question now facing us is two-fold: First, is Poincaré's criticism just? In other words, is Dedekind's mathematical continuum nothing more than discrete entities collected together? And second, if so, in what sense is this collection of the real numbers justly called a "continuum"? What role does such a continuum play in mathematics, in geometry, in calculus? In order to answer these questions, we must first examine Dedekind's theory of the nature of numbers in general, so that we may better understand his theory of real numbers in particular, and thus understand what it may mean for these real numbers to be called continuous.

---

[19]Henri Poincaré, as quoted in Russell [1903, p. 347].

3.3. Dedekind's Theory of Number

Dedekind's seminal work on the nature of numbers, *Was sind und was sollen die Zahlen?*[20] was written over twenty years after *Stetigkeit und irrationale Zahlen*, though Dedekind took it to be merely an elaboration of the arguments forwarded in *Stetigkeit*. In *Was sind und was sollen die Zahlen?* he develops the notion of a *chain*, establishes the existence of *simply infinite* chains, and uses these two ideas as a logical foundation for the major tenets of arithmetic. This course of development is remarkably similar to other nineteenth century projects, such as that of Guiseppe Peano (1858–1932); in fact, Peano himself was inspired by Dedekind's essay. Although Dedekind did not axiomatize his system as explicitly as Peano did, it is organized in such a manner that axioms can essentially be culled from his exposition.

Essential to this development is Dedekind's characterization of number, which is based on a definition of the natural numbers in particular.

> If in the consideration of a simply infinite system N set in order by a transformation φ we entirely neglect the special character of the elements; simply retaining their distinguishability and taking into account only the relations to one another in which they are placed by the order-setting transformation φ, then are these elements called *natural numbers* or *ordinal numbers* or simply *numbers*.[21]

Loosely speaking, the claim here is that that if a system is ordered in a certain way, if we can simply ignore all unique features of the elements, the resulting generalization will be the system of natural numbers, and each generalized element will itself be a number. Before we define each aspect of this definition, first note that Dedekind does not focus on some mysterious property held within each individual natural number – the "twoness" contained within the number two, or any such nonsense. What gives a natural number its character is its ability to be distinguished from other numbers, and the relations of order it has to other numbers.

---

[20]Translated as "The Nature and Meaning of Numbers;" page numbers refer to the *Essays on the Theory of Numbers* volume.
[21]See Dedekind [1963, p. 68].

The concept of *element* is likewise as free of content as possible. The elements of Dedekind's systems, the things that gain their numberhood from being organized in this fashion, are any things we can think of; literally they are "every object of our thought."[22] Anything thinkable counts as a possible element of one of these systems, and thus counts as a possible number, as long it can be organized into a simply infinite system. Not only does the number two not have some particular sense of two-ness, it does not have a particular numerical property at all; it is simply an object of thought bearing the proper relationship to other objects of thought in the proper type of system.

To understand the proper type of system, that is, a *simply infinite system*, first one must understand Dedekind's *transformation*, and his *chain*. A transformation is a law which assigns every element in a system to a determinate thing, which may or may not itself be a member of the same system.[23] A chain is a system K such that there is a transform K' of K that is a part of K itself. For example, take each of the natural numbers n, and apply the $2n$ transform. It is clear that $2n$ is a transformation, as every element in the system of natural numbers is assigned to a determinate thing, *viz.*, to a particular even number. The result of collecting all of these determinate things into their own system would give us the even numbers, which are themselves part of the system of natural numbers. Thus, $2n$ is a transformation which produces a chain. In modern set theory, a chain is any set such that there is a function mapping the set into a subset of itself.

A *similar transformation* is any two-way transformation, i.e., any one-to-one correspondence. Dedekind then defined an infinite system as any system which can be similarly transformed into a proper part of itself.[24] This is notable because, as was discussed in Chapter 1, for centuries the fact that the natural numbers could be mapped onto a proper part of themselves was considered a paradox, sometimes called "the paradox of different infinites." In fact, merely forty years before the publication of Dedekind's *Was sind und was sollen die Zahlen?*, this feature of infinite sets was still viewed as troublesome. Bernard Bolzano (1741–1848) noted that while there are infinitely many rational numbers between zero and five, and infinitely many between zero and twelve, the

---

[22]Ibid., p. 44.
[23]Ibid., p. 50.
[24]Ibid., p. 63.

3.3. DEDEKIND'S THEORY OF NUMBER 47

two sets could be put into one-to-one correspondence, though "the latter set [is] greater than the former, seeing that the former constitutes a mere part of the latter."[25] Bolzano wrote of this correspondence, "As I am far from denying, an air of paradox clings to these assertions," but he solved the paradox by insisting that the one-to-one correspondence only proved that if one set was infinite, they both must be, and that one-to-one correspondence only established equimultiplicity in the case of finite sets. Thus, for Bolzano, two such sets are in fact both infinite, but not equally large.[26] Dedekind retained the assertion that one-to-one correspondence established *equimultiplicity,* but abandoned the belief that relative size can be determined by the part/whole relationship. Rather than viewing this one-to-one correspondence between a collection and a proper part of that collection as a paradox, Dedekind uses this condition as the very definition of an infinite collection.

Having defined chains and transformations, we can now define *simply infinite systems*:

> A system N is said to be *simply infinite* when there exists a similar transformation φ of N in itself such that N appears as a chain of an element not contained in $\varphi(N)$.[27]

A simply infinite system is thus a system which has a transform that maps the system into a chain which itself contains the entire system, except for one element. The element lacking from this chain is the first element the transform is applied to, the 'base element' of the chain. Succession is the perfect example of a similar transformation leading to a simply infinite system, but many other simply infinite systems are possible, based on many different similar transformations.

For a somewhat more humorous example of a simply infinite system, consider the (possibly apocryphal) story of Bertrand Russell faced with a stubborn elderly woman at one of his lectures. Reportedly, the woman argued that the

---

[25]Bolzano, *Paradoxes of the Infinite,* as quoted in Waldegg [2005, p. 569].

[26]Ibid., p. 570. It is interesting to note that Peirce had a rather similar attitude, as displayed by his often repeated "syllogism of transposed quantity," as we will see in Chapter 6.

[27]See Dedekind [1963, p. 67].

world was supported by a large turtle. When Russell asked what the turtle was standing on, she replied "another turtle." After Russell asked what that turtle was standing on, she is supposed to have replied, "You can't fool me, young man! It's turtles all the way down."[28] Supposing the woman were correct, this infinite regression of turtles would itself form a simply infinite system, with "x rests on the back of y" as the similar transformation. The collection of every such x is a simply infinite system; the collection of every y forms a chain that includes everything in the system but the base element – in the case of the turtle example, the earth itself.

With the key terminology in place, we are now in a position to return to Dedekind's full definition of number. To review the terminology, we began with *elements*, which are any object of our thought, and an element is "completely determined by all that can be affirmed or thought concerning it."[29] Elements can be collected into *systems*, some of which have particular characteristics. Considering systems that are *simply infinite*, we note that all simply infinite systems are similar to one another (isomorphic), being organized in the same manner. Next,

> We entirely neglect the special character of the elements; simply retaining their distinguishability and taking into account only the relations to one another in which they are placed by the order-setting transformation φ.[30]

The things, remember, were simply objects of our thought. We ignore any irrelevant characteristics of these objects, such as the color or the size of the object of thought, and focus merely on their distinguishability and their relationship to one another. These elements, in these systems, simply *are* the natural numbers. Dedekind referred to the act of stripping away irrelevant qualities from the elements as "freeing the elements from every other content (abstraction)."[31]

---

[28] Stephen Hawking opens his book, *A Brief History of Time*, with a slightly different form of this anecdote.

[29] See Dedekind [1963, p. 44].

[30] Ibid., p. 68.

[31] Ibid., p. 68.

Given that we began with random objects of thought, the question must arise, has Dedekind defined one system of natural numbers, or instead many different ones? While it is perfectly possible that there are many such simply infinite systems that fit the appropriate requirements, one must assume that his method of abstraction saves us from needless duplication of number systems. Consider again the system of infinitely regressing turtles. Turtles are objects of our thought, as is the earth, and these objects can be mentally arranged into a simply infinite system. Applying Dedekind's method of abstraction would mean we could no longer distinguish between a turtle and the earth, but only between the base element, the next element, etc. It also would mean that we can no longer distinguish between turtles in an infinite regress and, say, notes in an infinite melody, as the features of turtles that distinguish them from notes would be abstracted away as irrelevant to their organization in the system and to their distinguishability from each other within the system.

The relations between these elements "are always the same in all ordered simply infinite systems, whatever names may happen to be given to the individual elements."[32] Therefore, we can abstract away from any names the elements bear naturally and name the elements of these systems what we will – calling the base element 1, as Dedekind does. Hence, the initially many isomorphic systems are thus abstracted, one from the next, so that in essence they collapse into one, which we can then call the number system.

The only possible hindrance to this tidy collapse into one system is when we consider the transformation φ. Nowhere does Dedekind specify that the nature of φ itself is abstracted away, and this makes a difference. Recall, in the case of turtles, the transformation is one of regression, while with notes in a melody, it is one of succession. There are differences in meaning between succession and regression, but also logical differences: one transform runs forward, while the other runs backward. However, the important feature of φ for Dedekind was that it gives an infinite system some order, and thus, one can possibly argue that we can abstract away from φ all of the peculiarities other than the sheer fact that it sets the elements of our system into an order.

---

[32]Ibid., p. 68.

Dedekind gives an example of a system thus organized, before abstracting away from all its particularities; in fact, he gives this example as proof of the existence of infinite systems themselves:

> My own realm of thoughts, i.e., the totality S of all things which can be objects of my thought, is infinite. For if s signifies an element of S, then is the thought s', that s can be an object of my thought, itself an element of S.[33]

Thus, S is the collection of all the things Dedekind could think, and the transform φ would be "x is an object of my thought." Beginning with a random thought s, we plug it into φ and are rewarded with

$$s' = \text{"s is an object of my thought"}$$

A similar transform can be made on s' itself, yielding s'', and so on, creating a simply infinite chain, the base of which is s itself. Abstracting away from the particulars linking this system to Richard Dedekind, to *his* thoughts in particular, and to the transform φ, it is obvious that the system is organizationally isomorphic to any other system we could organize in this manner, and thus, according to Dedekind, the general features such systems have in common are the numbers themselves.

After thus establishing the system of natural numbers as any simply infinite system with its particularities abstracted away from it, Dedekind then demonstrates how the rational numbers can be created from the natural numbers. Or rather, he demonstrates how first the operations typically performed on natural numbers, and then the rational numbers themselves, follow with "natural consequence" from the naturals:

---

[33]Ibid., p. 64. Interestingly, Bolzano uses a very similar argument to establish the existence of actual infinities. However, rather than an object of thought serving as the base case and "can be thought" as the transformation, Bolzano uses "any truth taken at random" as the base case, calls it A, and notes that the proposition "A is true" is distinct from A itself, forming B, from which we can form $C = $ "B is true", etc. To complete the proof that the set so formed is in fact infinite, Bolzano subjects it to what he considers the ultimate proof of infinity: he shows it is in one-to-one correspondence with the set of natural numbers. As Dedekind is here attempting to establish the natural numbers with logical rigor, he must refrain from using this last step to avoid circularity. See Waldegg [2005, p. 567].

3.3. DEDEKIND'S THEORY OF NUMBER

> I regard the whole of arithmetic as a necessary, or at least natural, consequence of the simplest arithmetic act, that of counting. [...] Addition is the combination of any arbitrary repetitions of [counting]; in a similar way arises multiplication. [...] Thus, negative and fractional numbers have been created by the human mind, and in the system of all rational numbers there has been gained an instrument of infinitely greater perfection.[34]

After the naturals are produced, all else follows logically; i.e., arithmetic, algebra, and even analysis immediately results "from the laws of thought."[35] So let us follow this progression. After counting, which, as we have seen, is logically founded by chain theory, one naturally wishes to perform addition – to speed up counting, so to speak. The combination of acts of addition gives us multiplication. Wishing to decrease as well as increase, one inverts addition to get subtraction. This act, however, is incomplete within the natural numbers themselves, if we wish to apply subtraction to any two numbers in our range; thus, negative numbers are needed to close the operation.

Similarly, the inverse of multiplication leads to the need for the creation of fractions, and thus the expansion of the system of numbers from the natural numbers to the rationals.[36] One might then take exponents as the combination of acts of multiplication, and the inverse operation to exponentiation as the creator of irrational numbers, but here we reach a problem. Taking the logical progression to the creation of the irrationals does not yield *all* of the irrationals: not all real numbers are reachable through taking the nth root of rational numbers. Nice, orderly irrationals such as the square root of 2 appear in this progression, but wildcards such as π do not. This is one reason Dedekind does not establish the irrationals through similar "operational" methods; but rather, establishes them through the method of cuts.

---

[34] See Dedekind [1963, p. 4].

[35] Ibid., p. 31.

[36] It is interesting to compare this 'just-so story' of the creation of numbers with the actual history of the inclusion of various mathematical entities in our numerical canon, particularly the rapid inclusion of negative numbers, fractions, irrationals, and imaginaries, all in the sixteenth century. Recall the brief discussion of this explosion of the number system in Chapter 2.

It was clear, through our progression from the naturals to the rationals, that the members of our new systems of numbers were a similar sort of thing as the members of the naturals; they completed operations we wished to perform on naturals, and fulfilled basic mathematical needs. Even the rationals, for example, were totally-ordered, and they were derived from the naturals using a simple operation. The metaphysical status of the new irrational numbers, however, is distinct. Our exposition reveals their nature: they are in essence gap-fillers. Look at the rationals, find a gap, then fill it in; call the filling an irrational number. It seems that the end result is a number system with different types of numbers – those directly derived from natural numbers by various operations, and those which fill the gaps between these derived numbers.

Now that we have a clear idea of what number means to Dedekind, we can investigate his continuity and answer the following questions: whether Dedekind's continuum is compositional, whether it is composed of discrete elements, and what role Dedekind's continuum plays in calculus itself.

3.4. The Nature of Dedekind Continuity

To review briefly, Dedekind defined "natural number," or simply "number," as the abstraction of any simply infinite system – any system that was sufficiently organized. From the naturals, he follows a logical progression to generate the integers and the rationals, and with cuts, he generates the reals from the rationals. The reals satisfy Dedekind's principle of continuity, and thus, he judges the collection of reals a continuum. Yet, looking over this progression of creation from the naturals to the reals, one thing is clear; Dedekind's continuum is distinctly compositional in nature. In Chapter 2 we reviewed Aristotle's proof that a continuum cannot be composed of indivisibles, and cannot be divided into them either; while Dedekind's numbers are not indivisibles in the classic Aristotelian sense, his continuum is composed of individual elements, and nothing more.

The inherent nature of these elements varies: the elements of the natural numbers are any thinkable object, organized and stripped of any particular features other than this organization itself; the rationals are ratios of naturals; the irrationals are cuts on the rationals. Yet, each number is an individual element, and Dedekind's principle of continuity does not posit essential connections between the members of the continuity; the principle only specifies that in a continuous entity, it is possible to make a cut between elements. Dedekind's reals are continuous, according to Dedekind, because the collection is full enough that no gaps can be found, but the collection is still a collection of elements.

Dedekind is the only one of our four mathematicians who was satisfied with such a strictly compositional continuum. As we shall see in later chapters, Cantor believed that in addition to a completeness of this sort, the elements of a continuum must also have an essential connection between them. Du Bois-Reymond attacked the idea of compositional continua such as this "on the ground that it was committed to the reduction of the continuous to the discrete, a program whose philosophical cogency, and even logical consistency, had been challenged many times over the centuries."[37] Peirce came to believe that compositional continua failed to satisfy our basic intuitions about the nature of continuity.

As was briefly noted in Chapter 2, Aristotle's argument against compositional continua does not succeed against Dedekind's principle of continuity. The thrust of Aristotle's argument was that compositionality required two elements to be next to each other (which he proves to be impossible), because if they were not next to each other, there would either be a gap between them (and thus a gap in the continuum), or there would be something foreign between them, and thus the continuum would not be composed entirely of these elements. Yet in Dedekind's system of real numbers, no two real numbers are next to each other, and yet there is no gap, and there is nothing foreign in the set; rather, there are always infinitely many real numbers between any two. Aristotle's argument thus does not refute Dedekind's principle of continuity; however, many philosophers and mathematicians believe that a continuum can never be composed of discrete elements. The question before us can thus be framed: are the numbers composing Dedekind's continuum *discrete* entities, and if so, does this prevent them from being viewed as continuous?

In 1917, Edward V. Huntington (1874–1952) argues that Dedekind and Cantor's real numbers form non-discrete series.[38] Huntington calls a *discrete*

---

[37]See Ehrlich [1994, p. x].
[38]See Huntington [1917].

series one which is (i) divisible into two parts K_1 and K_2, such that every element of K_1 precedes every element of K_2, and that there is at least one x such that any element preceding x belongs to K_1 and every element following x belongs to K_2; and (ii) every element of the series except the last has an immediate successor, and every element of the series except the first has an immediate predecessor.[39] Thus, the integers are a good example of a discrete series, but the reals, failing the second condition, are not. Notably, the most important feature of a discrete series for Huntington is that one can perform induction on the series. However, Huntington himself never refers to the *elements* of any series as discrete or non-discrete, only to the series themselves as discrete or not. Thus, while Huntington quite clearly states that the real numbers form a continuous series, and hence *not* a discrete series, nowhere does he state that therefore the real numbers are not discrete elements.

Conversely, John L. Bell argues that all numbers are discrete, and also that discreteness is commonly regarded as antithetical to continuous. First, whole numbers are the essence of discreteness: "In mathematics it is the concept of whole number, later elaborated into the set concept, that provides an embodiment of the idea of pure discreteness, that is, of the idea of a collection of separate individual objects."[40] Geometry, on the other hand, is the essence of continuity: "by their very nature geometric figures are continuous; discreteness is injected into geometry, the realm of the continuous, through the concept of a point, that is, a discrete entity marking the boundary of a line."[41] The difficulty with discrete numbers being used to analyze the continua of geometry, Bell says, arose as early as Pythagoras and his discovery of incommensurable magnitudes. "Here the realm of continuous geometric magnitudes resisted the Pythagorean attempt to reduce it to the discrete form of number."[42]

Bell separates late nineteenth century mathematicians into two categories: those who believe the continuous is not reducible to discrete entities, such as du Bois-Reymond, Giuseppe Veronese (1854–1917), Hermann Weyl (1885–1955), L. E. J. Brouwer (1881–1966), and Peirce; and those who not only believed in the reducibility of the continuous to the discrete, but in some cases, believed they

---

[39]Ibid., p. 19.
[40]See Bell, p. 1.
[41]Ibid., p. 1–2.
[42]Ibid., p. 2.

3.4. THE NATURE OF DEDEKIND CONTINUITY

had accomplished just such a reduction, like Dedekind and Cantor. Thus, a full analysis of whether Dedekind's reduction of the continuous to the discrete by means of cuts actually works can come only after we have examined the position and arguments of the other figures to be considered in this dissertation: viz. Cantor, du Bois-Reymond, and Peirce.

Pending the fuller analysis, we can still judge Dedekind's continuum on his own stated goals. Dedekind wished to create this real number continuum to free calculus from necessary geometrical references. In this, he succeeds: he has defined a concept of continuity which is completely arithmetical; a function can now approach a limit continuously, numerically, consistently. No tangent lines or curves are necessary for this analysis. However, one might ask if Dedekind's separation between geometry and systems of numbers has now gone too far. In terms of logical foundations, it was necessary to free calculus from geometrical references; in terms of application, however, it is necessary to be able to apply calculus to realms outside of algebra and number theory. Thus, now that Dedekind has succeeded in shoring up the algebraic foundations of calculus, one might ask if it is still possible to apply this algebraic calculus to space and time.

One of the greatest difficulties in this application seems to stem from Dedekind's insistence that mathematics is both mental, and logical. Thus:

> In speaking of arithmetic (algebra, analysis) as a part of logic I mean to imply that I consider the number-concept entirely independent of the notions or intuitions of space and time, that I consider it an immediate result from the laws of thought.[43]

And again:

> Without any notion of measurable quantities and simply by a finite system of simple thought-steps man can advance to the creation of the pure continuous number-domain.[44]

Thus, if numbers are thought and logic, and Dedekind specifically denies any reference to measurable quantities or anything non-mental, how can we trust

---
[43]See Dedekind [1963, p. 31].
[44]Ibid., p. 38.

that the resulting system applies to our physics, or even to our carpentry? Further mysteries arise when, after proving that his real numbers form a continuous system, Dedekind suggests that space itself may not be continuous:

> For a great part of the science of space the continuity of its configurations is not even a necessary condition. [...] If any one should say that we cannot conceive of space as anything else than continuous, I should venture to doubt it.[45]

Thus, if Dedekind's real number system is necessarily continuous, but space might not be, in what sense do they link up?

Dedekind did believe that space was in fact continuous, and it is clear that he believed that geometry gives us our paradigm of what continuity itself. However given that he divorced numbers from geometry in the development of the calculus, their reunion becomes something that is not quite so obvious. Perhaps we can reunite them along these lines: given that cuts assure us that the real numbers have no gaps, one will no longer run into the same difficulties of incommensurability as the Greeks did. That is, wherever we need to measure, there will be a number there. We will not reach a situation where the reals are not sufficient to measure a triangle, as the counting numbers proved to be. Thus, while Dedekind has not proved a correspondence between the points on a line and the real numbers, the assumption of such a correspondence is not immediately refuted by incommensurable quantities.

3.5. The Relationship Between Dedekind's Continuity and Infinitesimals

Before we end our discussion of Dedekind and turn to Cantor, we can draw some interesting conclusions about infinitesimals from Dedekind's system of cuts itself. Though Dedekind did not discuss infinitesimals directly, his real number system creates a calculus that does not need to rely upon infinitesimals, and his conception of continuity mathematically excludes infinitesimals. One of the most intriguing features of an infinitesimal quantity is that adding an infinitesimal to a finite quantity such as 1 simply results in that same finite

---

[45]Ibid., p. 37–38.

3.5. DEDEKIND'S CONTINUITY AND INFINITESIMALS

quantity; the quantity has not been moved, even by a very small amount. Therefore, as Cantor points out (in an argument we will examine more closely in the next chapter), infinitesimals violate the Archimedean principle. The Archimedean principle states that "if a and b are any two positive numbers of [a] system and $a < b$, then it should be possible to add a often enough that the sum $a + a + \cdots + a$ eventually surpasses b. Briefly stated, there should always exist a natural number n such that $na > b$."[46] It is clear that if infinitesimals are considered to be numbers, they must be non-Archimedean numbers; if a in the above definition is an infinitesimal, no natural number n, no matter how large,[47] can take a past another number b.

Thus, infinitesimals are excluded in any Archimedean system, but Dedekind's continuity logically implies the Archimedean principle. This can be proved as follows.

THEOREM. *If any cut of the reals is determined by a unique real number (the assumption of Dedekind continuity), then for any positive reals a and b, there is a natural number n such that $na \geq b$.*

PROOF. Assume for *reductio* that there exist two positive reals a and b such that $na < b$ for any natural n.

From the series $a, 2a, 3a, \ldots$ we can form the set
$$A = \{x \mid x \leq na, \text{for any natural number } n\}.$$
It is clear that b lies outside of A, and since b is a real number, there are infinitely many numbers greater than b which also lie outside A. In fact, we can form another set
$$B = \{x \mid x > na\}.$$
A and B together encompass all the reals, and it is clear that every member of A is less than every member of B, and thus (A, B) constitutes a cut. By our original assumption, (A, B) must be determined by a unique real number; call it c. Now, c must be the greatest member of A or the least member of B.

Assume c is the greatest member of A. Thus, by definition of A, $c \leq m$ for some m. But if $c < ma$, then since ma is itself a member of A (since A contains

---

[46]See Waismann [2003, p. 209].

[47]And, Cantor says, no transfinite number either. More on that in Chapter 4.

all $\{x \mid x \le na\}$ for all n), c would not be the greatest member of A, unless $c = ma$. Therefore, $c = ma$. Yet this cannot be the case either, for if $c = ma$, then since $(m + 1)a > ma$, by substitution, $(m + 1)a > c$. But $(m + 1)a$ is a member of A (by definition of A, since $(m + 1)$ is a natural number), and thus c would still not be the greatest member of A.

Therefore, c must be the least member of B. By definition of B, $c > na$ for all n. Notice also, that for every real number $r < c$, then r is a member of A (since every real is either in A or B, and as c is the least member of B, for all x member of B, $x \ge c$).

Consider, then, $(c-a)$. As c and a are both positive real numbers, $(c-a) < c$, and therefore, $(c - a)$ is a member of A. By definition of A, $(c - a) \le ma$ for some natural m.

But if $(c - a) \le ma$, then $c \le ma + a$, which is equivalent to $c \le (m + 1)a$. However, $m + 1$ is a natural number, and thus by the definition of A, c must be a member of A, which contradicts the above. $\square$

Thus, the existence of infinitesimals is inconsistent with Dedekind continuity. Since Dedekind has become so influential, several proponents of infinitesimals have tried to work with modified versions of Dedekind-continuity – versions which retain the main advantages of the system without also insisting on the Archimedean principle. However, as Dedekind defined continuity, though he wrote nothing in these pages directly against infinitesimals, his system is confirmedly non-Archimedean.

We now have examined several features of Dedekind's continuity. (i) The most important feature of a continuous collection is that it cannot contain gaps. (ii) The real numbers, when defined as cuts on the rationals, exhibit this principle of continuity. (iii) Dedekind's continuity, given his theory of number, is thus compositional in nature. (iv) Infinitesimals are inconsistent with Dedekind's principle of continuity, as the latter implies the Archimedean principle.

As we shall see in the next chapter, Cantor created a definition of continuity that shared some similarities with Dedekind's, but attempted to overcome some of the problems Cantor saw in Dedekind's theory. Cantor similarly saw the real numbers as created from the rationals, and saw the real numbers as continuous (and in fact as exhibiting the very essence of continuity itself), but he wished to overcome the charge of his continuity being a simple collection of discrete elements by mathematically defining an essential connection between the elements.

CHAPTER 4

Georg Cantor

4.1. Biography and Introduction

Georg Ferdinand Ludwig Philipp Cantor was born in St. Petersburg, Russia, in 1845, the son of a relatively wealthy merchant. Cantor was taken to Germany at age 11, and was sent to study at a trade school in Darmstadt at age 15, in preparation for a career in engineering. When he finished at Darmstadt in 1862, he decided to commit himself to a career in mathematics. As Dauben wrote, "From the very beginning, apparently, Cantor had felt some inner compulsion to study mathematics."[1] He began his mathematical education at the Polytechnic of Zurich, though his studies were cut short by his father's death in 1963. He moved to the University of Berlin, attending lectures by Karl Weierstrass (1815–1897) and Leopold Kronecker (1823–1891), and completing his dissertation on number theory in 1867.

Cantor's career was a vigorous one, his brilliance earning him acclaim (such as his promotion to Extraordinary Professor at Halle in 1872, a mere three years after he was appointed there), his innovation and enthusiasm earning him enemies and embroiling him in mathematical controversies. Among the most controversial aspects of his work, and also among the most influential, must be numbered his set theory and his transfinite number theory. Though this chapter will not deal directly with these two theories, they are not unrelated to our project here; his set theory was developed in conjunction with his real

---
[1] See Dauben [1990, p. 277].

number theory, and his investigations of mathematical continuity and infinity played no small part in his development of transfinite theory.[2]

Though Cantor's real-number theory and treatment of mathematical continuity are in many ways similar to those of Dedekind, Cantor goes beyond Dedekind in significant ways, and thus a careful treatment of Cantor's thoughts and developments in this arena is both interesting in itself and necessary for a full understanding of Cantor-Dedekind continuity. As Cantor often developed his mathematical theory as needed to prove the particular theorems on which he was working, the best means of understanding his theory of the mathematical continuum is to follow his mathematical progression toward his eventual definition of continuity. Thus, this chapter will proceed chronologically. Section 4.2 will discuss Cantor's 1872 work on trigonometric series and the development of his real-number theory, Section 4.3 will treat the papers in which he develops nondenumerability theory, from 1872 to 1878, Section 4.4 discusses his early theory of continuity, as it appears in an 1878 paper, and Section 4.5 presents Cantor's most thorough and philosophical discussion of continuity, contained in the 1883 work primarily concerned with a defense of transfinite theory. Finally, in Section 4.6, we will consider in some detail his famous argument against the existence of infinitesimals, which he believed were self-contradictory entities.

4.2. Real Numbers (1872)

Cantor developed his theory of real numbers in the 1872 article, "Über die Ausdehnung eins Satzes aus der Theorie der trigonometrischen Reihen."[3] This essay is one in a series of Cantor's early papers (1870–1872) which were concerned with particularly interesting trigonometrical series. As we shall see below, Cantor developed his theory of real numbers as a tool to prove the uniqueness theorem he was working on in 1872; thus, a brief overview of Cantor's project would not be out of place.

---

[2]For a more thorough biography of Georg Cantor, see [Dauben, 1990, p. 277], especially Chapter 12, as well as the University of St. Andrew's biography of Cantor, [Online]www-groups.dcs.st-and.ac.uk/history/Biographies/Cantor.html.

[3]See Cantor [1872]. Translations from the French are the present author's, unless otherwise noted.

4.2. REAL NUMBERS (1872)

This uniqueness project has its roots in the early nineteenth century, with the introduction of Fourier series. Joseph Fourier (1768–1830), while studying the conductivity of heat, "had established that arbitrarily given functions could be represented by trigonometric series with coefficients of a specified type."[4] The term "Fourier series" refers to just such expansions of functions as trigonometric series, and these series are a powerful mathematical tool used in a variety of equations, proofs, and sciences. Many mathematicians throughout the nineteenth century spent considerable time expanding on the notion of Fourier series, making the transformations more general and applicable to a wider variety of functions, generally making the tool more powerful.

Cantor's 1872 paper is an attempt to establish uniqueness conditions for one such transformation from certain functions to trigonometric series; i.e., he was working on the problem of whether an arbitrary function could be represented by exactly one trigonometric expansion. As Cantor wrote, "I would like to make known in this work an extension of a theorem according to which a function can be developed by a trigonometric series using exactly one method."[5] He had earlier established uniqueness conditions under certain restrictions, when he proved the theorem,

> That two trigonometric series;
> $$1/2 \, b_0 + (\Sigma a_n \sin nx + b_n \cos nx)$$
> and
> $$1/2 \, b_0^1 + (\Sigma_n^1 \sin nx + b_n^1 \cos nx)$$
> which, for all values of x, converge and have the same sum, have the same coefficients; I further showed ... that this theory stays true if, for a finite number of values x, one renounces either the convergence, or the equality of the sums of these two series.[6]

The purpose of the 1872 paper is to extend this result.

> The extension I have in view here is this: that for an *infinite* number of values of x in the interval $[0, (2\pi)]$ one can renounce

---

[4]See Dauben [1990, p. 6].
[5]See Cantor [1872, p. 336].
[6]Ibid., p. 336. Cantor originally established this theorem in Cantor [1870, p. 139].

either the convergence or the agreement of the sums of the series, without the theorem ceasing to be true.[7]

Thus, Cantor's project in 1872 was to expand his uniqueness results by extending the range of values under which the conditions placed on uniqueness can be relaxed.

He does prove this extension in the last section of the paper, but the details of this proof are beyond the scope of our considerations; rather, our focus is on the tool Cantor needs to make the jump from finitude to limited (bounded) infinity – that is, a precise mathematical understanding of the real numbers. Thus, Cantor developed a theory of real numbers, and did so in a manner somewhat similar to that of Dedekind – by using rational numbers as the foundation. Cantor's rationals included zero, and the set of rationals is called A.

His construction of irrational numbers begins with an infinite series of rationals. This infinite series must be "obtained by a law"[8] and is represented thus:

$$a_1, a_2, \ldots a_n, \ldots$$

The series is constituted such that the difference $a_{n+m} - a_n$ becomes "infinitely smaller as n grows."[9] Here, n is an arbitrary integer such that $(a_{n+m} - a_n) < \epsilon$, where ϵ is a positive rational and m is an arbitrary integer. In later papers, Cantor called this type of series a "fundamental sequence," which is defined more concisely in Dauben:

> The infinite sequence
>
> $$a_1, a_2, \ldots, a_n, \ldots$$
>
> is said to be a fundamental sequence if there exists an integer N such that for any positive, rational value of ϵ, $|a_{n+m} - a_n| < \epsilon$, for whatever m and for all $n > N$.[10]

---

[7]Cantor 1872, p. 336–337.
[8]Ibid., p. 337.
[9]Ibid., p. 337.
[10]Dauben, p. 38. Though Cantor does not refer to them as such, these fundamental sequences are usually referred to by the name that Bolzano gave them, "Cauchy sequences."

4.2. REAL NUMBERS (1872)

Notice that while these two definitions pick out the same sequences, the emphasis in each one is different. Cantor's original definition draws attention to an important feature of fundamental sequences – that the members of the sequence draw ever closer to each other as the sequence progresses whereas Dauben's version focuses on the strict containment of the differences between a_{n+m} and a_n.

Such sequences had long been used to define sets that had limits without assuming the existence of the limit itself, which is, no doubt, why Cantor chose to use them to define his real numbers. For, as he wrote in 1883, the main mistake encountered in previous attempts to establish the real numbers based on some collection of rationals is the assumption of the existence of the very limit used to define the real number. Cantor wrote:

> I believe that this logical mistake, which was first avoided by Herr Weierstrass, was committed almost universally in previous times, and not noticed because it belongs among those rare cases in which actual mistakes cannot cause any significant damage to the calculus.[11]

The important point, which Cantor noticed, is that such sequences are a natural tool to use if one wishes to develop real numbers as limits without previously assuming their existence.

After thus defining fundamental sequences, Cantor associates a "determined limit b"[12] with each sequence. These determined limits will eventually be established as irrational numbers, but for the moment, they are merely symbols associated with particular sequences. As Dauben writes, the phrase "determined limit" must be understood "as a convention to express, not that the sequence $\{a_n\}$ actually had the limit b, or that b was presupposed as the limit, but merely that with each such sequence $\{a_n\}$ a definite symbol b was associated with it."[13]

---

[11]See Parpart [1976, p.81], the English translation of Cantor's *Grundlagen*, [Cantor, 1883].

[12]Cantor 1872, p. 338.

[13]Dauben, p. 38. The original reads "[...] a definite symbol a was associated with it," but it seems obvious b is intended here.

Cantor then proves that such b's associated with such sequences are linearly ordered. Let $\{a_n\}$ and $\{a'_n\}$ be two non-equal fundamental sequences, with b and b' their respective associated determined limits. One (and only one) of three relationships must hold between the two sequences. Either: (i) $a_n - a'_n$ becomes infinitely smaller as n grows, or (ii) $a_n - a'_n$, after an n of a certain size, stays always larger than a positive rational magnitude ϵ, or (iii) $a_n - a'_n$, after an n of a certain size, remains smaller than a negative rational $-\epsilon$.[14] If (i) is the case, then $b = b'$; if (ii) is the case, then $b > b'$; and if (iii) is the case, then $b < b'$. Thus, the sequences (under the equivalence relation) are linearly ordered, and so too are their associated limits.

Cantor next shows that the conjoined set of B (the set of all such b's) and A (the set of rational numbers) is itself linearly ordered, by showing the same three relationships hold between any member of A and any member of B. Finally, he then indicates that there is good reason to suppose that these symbols b, b', b'', and so forth are the actual limits of the sequences with which they are associated.

> From these definitions and from those which immediately follow, it results (and can be rigorously demonstrated) that, b being the limit of series (1), $b - a_n$ becomes infinitely smaller as n grows, which justifies consequentially in a precise manner the designation of "limit of series (1)" given to b.[15]

This is far from proof that b is an actual limit of its associated sequence; it is not even proof that the sequences in question have limits. Cantor's real numbers, at least as far as this article goes, are the fundamental sequences themselves; b or b' is simply an easy way to refer to the sequence, and it seems helpful if we can think of b as a limit of $\{a_n\}$.

Thus, after proving that the elements of his set B can satisfy the operations of addition, subtraction, multiplication and division, and after a small detour,[16]

---

[14]Cantor 1872, p. 338.

[15]Ibid., p. 338.

[16]The detour is an interesting one historically, if not directly relevant to our current thesis. After developing the set B (which is, recall, the collection of determined limits associated with fundamental sequences from set A), Cantor claims we can take fundamental sequences of members of B and use them to develop a set C, and reiterate the operation on C to produce

4.3. CONTINUITY AND DENUMERABILITY (1872 - 1878)   67

Cantor went on to demonstrate how this collection of b's could be used in the measurement of distances. This demonstration is key to our thesis here, as it will be used later in his definition of continuity. Thus, in the next section we will examine this demonstration. We will also discuss an important feature of the real numbers; i.e., their nondenumerability.

4.3. Continuity and Denumerability (1872 - 1878)

After Cantor developed the set B as described above, (and indicated how further sets, C, D, etc. could be generated), he began referring to the elements of these sets as numerical magnitudes rather than as symbols. That is, he believed himself to have established numbers of a sort, though perhaps a peculiar sort of number.

> In the theory I have here described (according to which numerical magnitude, having at first, in general, in itself, no objectivity, appears only as an element of theorems which have a certain objectivity, of this theorem, for example, that the numerical magnitude serves as the limit to a corresponding sequence) it is essential to maintain the abstract distinction between B and C.[17]

Thus, though Cantor treated these elements as numbers, he restrained himself from embracing their objectivity as numbers, instead accepting their tenuous nature as entities merely associated with particular sequences.

In order to have them function fully as numerical magnitudes, however, it is not enough to prove that they are linearly ordered and that they satisfy basic equations; he must also establish that they can be used in measurement. This is

---

D, etc. He here claims that while this process could be iterated indefinitely, the only time the process results in a jump in magnitude is the movement from set A to B. B is thus in one-to-one correspondence with all such sets save A itself. Thus, Cantor was thinking about and working with magnitudes of infinity as early as this 1872 paper – a process, of course, that will eventually culminate in transfinite theory.

[17]Ibid., p. 340.

the task he took up in Section 2 of his 1872 paper, and eventually accomplished by means of a connecting axiom.

> It now follows without difficulty that numerical magnitudes from systems $C, D, \ldots$ are also capable of determining known distances. But to achieve knowledge of the bond that we observe between the systems of defined numerical magnitudes in §1 and the geometrical straight line, we must still add an axiom, here enunciated simply: to each numerical magnitude reciprocally belongs a determined point of the line of which the coordinate is equal to this numerical magnitude in the sense discussed in §1.[18]

That the connection is assumed and not proved is no accident. Cantor continues:

> I call this theorem an axiom because it is in its nature that it cannot be proved in a general fashion.[19]

As we will see a little further on, Cantor does indeed take this unprovable connection as the essence of continuity itself, not just the essence of the real number continuum, thus its status as an assumption is a notable one. In Section 4.5, when we discuss Cantor's definition of continuity itself, it will be good to keep in mind that Cantor believes this connection between the real numbers and points on the geometrical line to be an essential but necessarily unproved connection.[20]

But first, since we are discussing Cantor chronologically, it is necessary to comment on denumerability, another important ingredient in his theory of the continuity of the real numbers. As is generally well-known, Cantor proved that the rational numbers can be put in one-to-one correspondence with the naturals, and are thus countable, or denumerable. The "power" of the naturals

---

[18]Ibid., p. 342.
[19]Ibid., p. 342.
[20]This axiom – that the real numbers are in one-to-one correspondence with the points of a geometrical line – is often referred to today as the Cantor-Dedekind axiom.

is thus the same as the power of the rationals, and Cantor calls the power of the naturals the smallest possible power of an infinite set.[21]

Obtaining a result remarkable even to Cantor himself, he next proved that, though the algebraic irrational numbers are also denumerable, the irrationals themselves, and thus the reals as a whole, are not. The jump from the denumerability of the rational numbers and the nondenumerability of the reals is not simply an interesting feature of the number system; it is often used as a partial justification for why the real numbers form a continuous set and the rational numbers do not. Certainly, as Dedekind showed, the real numbers have a sort of completeness: wherever one wishes to divide the set, one is always able to do so at a number. The nondenumerability of the real numbers adds a further intuitive reason to suppose them to be a good candidate for continuity. The reals have a larger power; they are more extensive than a denumerable set. Of course, nondenumerability alone does not suffice for continuity, but some do take it as evidence that the real numbers are in a separate metaphysical category from the rationals.

4.4. Early Real-Number Continuity (1878)

Returning to Cantor's theory of continuity, we move now to his essay of 1878, *Ein Beitrag zur Mannigfaltigkeitslehr*. In this essay, Cantor abandoned trigonometrical series, and was working directly with the implications of set theory and various theorems one could prove about the set of real numbers. Here he discussed continuity directly, analyzed the relationship between different continuous sets, and discussed the relationship between continuous and noncontinuous sets. The *Beitrag* thus gives us several important insights into his early theory of continuity.

The main point of the *Beitrag* is to prove the theorem

> (A) Let $x_1, x_2, \ldots, x_n$ be n real variable magnitudes, independent of each other, such that each variable can take every value ≥ 0 and ≤ 1. Let t be another variable with the same

---

[21]See Cantor [1878]. References to the *Beitrag* use the pagination of the French translation.

limits ($0 \leq t \leq 1$). We can make this magnitude t correspond to the system of n magnitudes $x_1, x_2, \ldots, x_n$ such that to each determined value of t belongs a system of determined values $x_1, x_2, \ldots, x_n$ and vice versa – to each system of determined values $x_1, x_2, \ldots, x_n$ belongs a certain value of t.[22]

(A) is derived straightforwardly from three theorems Cantor wished to establish for continuous sets. First, we can completely and uniquely relate a continuous set of n dimensions to a continuous set of one dimension. Second, the elements of an n-dimensional continuous set can be uniquely determined by a continuous and real coordinate t. Third, the elements of an n-dimensional continuous set can also be uniquely determined by a system of m continuous coordinates $t_1, t_2 \ldots, t_m$. With these three theorems, one can reduce continuous sets *of any dimension* to sets *of one dimension*.

After several intricately entwined lemmas and sub-proofs, Cantor finally arrived at the (A) he desired – though not in an uncontroversial manner.[23] Along the way, he investigated more thoroughly the axiom introduced in his 1872 paper and discussed in Section 4.3 above – the postulate that the numbers do indeed correspond to the geometrical line. It is noteworthy that Cantor, like Dedekind, developed his real numbers first and only then axiomatized their relationship to geometry. As Ferreiros wrote,

> It had been customary to assume that the continuity of space or of the basic domain of magnitudes induces, through the definition of real numbers as ratios, the continuity of the number system. But now we find two mathematicians who emphasize the point that it is possible to define abstractly a continuous number system, while geometrical space is not necessarily continuous. One needs an axiom, sometimes called the axiom of Cantor-Dedekind, to postulate that space is continuous.[24]

---

[22] See Cantor [1878, p. 315].

[23] Not surprisingly, Cantor's constant foe, Leopold Kronecker, led the critical charge against the arguments in this article. See Dauben [1990, p. 66–72] for a discussion of Kronecker's opposition to Cantor's theories in general, and of his objections to the theories discussed here.

[24] See Ferreriós [1993, p. 135].

Or, at the very least, one needs an axiom to link our abstractly defined real continuum with geometry, and hence, one supposes, with space. Our first observation, then, regarding the *Beitrag* and Cantor's continuity, is that he still was firmly of the belief that the relationship between the real numbers and points on a geometrical line must be assumed and not proven.

Another claim in the *Beitrag* which is interesting for us is that continuous sets must be non-denumerable. After discussing various properties of sets of the "first power" (i.e., sets that can be put into one-to-one correspondence with the natural numbers) Cantor introduced the notion of a continuous set *as* a non-denumerable set.

> We will now in what follows examine the sets that we call continuous from the point of view of their power. [...] [I]t is certain that these sets do not belong to the first class, i.e., they have a power superior to that of the first.[25]

Thus, "from the point of view of their power," continuous sets must be non-denumerable. After this point in time, mathematicians will frequently list non-denumerability as a necessary though not sufficient property of any continua, but it is particularly interesting that Cantor specifies nondenumerability as a necessary condition so soon after proving that the reals have this property. In the space of five years, nondenumerability rose in status from a startling and non-obvious quality of the reals to an essential quality of any continuum.

The last observation I wish to make on Cantor's continua in this 1878 paper concerns the dimension-reducing project itself. When Cantor claimed that a continuum of any dimension could be treated mathematically as a continuum of one dimension, i.e., as a straight line, he did so by arguing that the members of a continuous set of n dimensions can be put into one-to-one correspondence with the members of a continuous set of one dimension. The philosophical implication of this is that any continuum, whether a line, plane, or three-dimensional figure, contains nothing more than the sum of its points. For Cantor, any continuum is thus completely defined by specifying the members belonging to it as a continuous set. We shall see Cantor expand on this belief in Section 4.5 below.

---

[25]See Cantor [1878, p. 313].

Before finally assembling all of these ingredients into Cantor's explicit definition of continuity, I wish to note one element in the *Beitrag* of historical interest. Cantor ends his 1878 paper with an early version of his Continuum Hypothesis:

> By a procedure of induction, the description of which we shall not enter into here, we are led to the theory that the number of classes of sets obtained after this style of grouping [i.e., grouping them according to power] is a finite number, and is equal to two.[26]

The claim here is that if we group our infinite sets by their powers we shall come up with only two groups. Cantor identified these groups as, first, the sets sharing the power of the positive whole numbers, and second, those sharing the power of the real numbers between 0 and 1. He would later prove that there are infinitely many distinct magnitudes of infinity, but he would hold firmly to the belief that there were no magnitudes of infinity larger than that of the natural numbers but smaller than that of the real numbers, and his search for definite proof of this Continuum Hypothesis would continue for most of his life.[27]

4.5. Infinity, and the Definition of Continuity (1883)

After first constructing real numbers in 1872, subsequently showing that the set of real numbers is nondenumerable in 1873, and comparing the real number continuum to continua of many dimensions in 1878, Cantor finally gives us an explicit definition of continuity in his *Grundlagen* in 1883. The *Grundlagen* is one of Cantor's most explicitly philosophical works, and intentionally so. In the preface, he wrote that this essay was intended primarily for two groups, "for philosophers who have followed the development of mathematics up to the most recent period, and for mathematicians who are familiar with the most important older writings and more recent works in philosophy."[28] Cantor was

---

[26]See Cantor [1878, p. 327].

[27]In the twentieth century, it was proved by Kurt Gödel and Paul Cohen that the Continuum Hypothesis can neither be proved nor disproved using the standard axioms of set theory; the hypothesis is logically independent of the rest of the axioms.

[28]Cantor 1883, p. 70.

4.5. INFINITY, AND THE DEFINITION OF CONTINUITY (1883)

quite aware that his mathematics – particularly his treatment of infinity – was causing a philosophical stir, and he wrote the *Grundlagen* particularly to defend some of the metaphysical and epistemological assumptions behind the acceptance of actual infinities implied by his transfinite theory. The *Grundlagen* is a complicated and fascinating essay; here I shall focus on only a small part of it. In particular, I wish to focus on Cantor's discussion of different types of infinity, his rejection of infinitesimals, and finally, his necessary and sufficient conditions for continuity.

In addition to distinguishing various magnitudes of infinity, Cantor distinguished between two different mathematical concepts of the infinite. The first concept of infinity, and the one Cantor regarded as more historically common, is the potential infinity contained within the concept of variable magnitude, "either growing beyond all limits, or diminishing to an arbitrary smallness, always, however, remaining *finite*."[29] This type of infinity Cantor refers to as *das Uneigentlich-unendliche* – the non-genuine infinite.

Cantor himself based his transfinite theory, and much of the mathematics of his later years, on a different concept of infinity, the *Eigentlich-unendliches*, or genuine infinite.

> According to this concept, in the investigation of an analytic function of a complex variable, for example, it has become necessary and in fact common practice to imagine in the plane representing the complex variable a single point at infinity, i.e., an infinitely distant but determinate point.[30]

This genuine infinite, the assumption of a point at infinity or the comparison of two actually infinite magnitudes, is the infinity Cantor was most concerned with, the concept assumed by transfinite theory, and the concept that many people of his time found philosophically objectionable. The main point of the *Grundlagen* is to defend this genuine infinite.[31]

---

[29]Ibid., p. 70.

[30]Ibid., p. 70.

[31]Paul du Bois-Reymond posited a similar distinction between types of infinities. See du Bois-Reymond [1887, p. 72–73] and Chapter 5, below.

The primary philosophical objection to genuine infinity, and the one which Cantor most wished to refute, is an argument found in Descartes, Spinoza, Leibniz, and others: namely, that humans, as finite beings, can never comprehend infinity, since infinity comprises the absolute; thus, only an infinite mind can truly comprehend this infinity. Cantor's rebuttal is simple: there are levels of magnitude between finitude and a true absolute; all of Cantor's magnitudes fall in these medium levels, and in fact, all magnitudes short of God himself belong to these middle-range, comprehensible infinitudes. Though these levels of infinity are, by definition, non-finite, they are nevertheless definite and determined – they are not finite, but they are limited. Thus, wrote Cantor, "All things, whether finite or infinite, are *definite* and, with the exception of God, can be determined by the intellect."[32] It is the absolute nature of God's infinity that Cantor agreed was incomprehensible by human beings; regular infinity was always bounded by something or another, always had limits and lacks, and thus was comprehensible.

Just as infinity is comprehensible by humans, so too is continuity, according to Cantor. He argued that comprehension of continuity did not require intuitive understanding of continuous entities, such as space or time.

> In my opinion, the enlistment of the *concept of* time or of the *intuition of* time in the discussion of the much more fundamental and more general concept of the continuum is not in order. It is my judgment that time is a notion which for its clear explication presupposes the concept of continuity, which is independent of it.[33]

The same holds true of space. An understanding of continuity, Cantor argued, does not depend upon a prior intuition of a continuum such as that of space or time. Rather, the opposite holds; understanding space or time requires a prior understanding of continuity itself, and understanding continuity itself requires "sober and exact mathematical investigations."[34]

---

[32]See Cantor [1883, p. 76].
[33]Ibid., p. 85.
[34]Ibid., p. 85.

4.5. INFINITY, AND THE DEFINITION OF CONTINUITY (1883)

Cantor's sober and exact investigations lead him to define continuous sets as having two conditions, which are individually necessary and jointly sufficient: a continuous set must be *connected*, and *perfect*. A set P is connected when between any two numbers t and t' at least one finite collection of fellow members $\{t_\nu\}$ could be found such that the distance between t_ν and $t_{\nu-1}$ are collectively smaller than ϵ, an arbitrarily chosen positive number.[35] From connectivity, everywhere-denseness follows.

A set P is perfect when it equals each of its derived sets $P^{(\gamma)}$. Cantor used the concept of derived sets in his 1872 paper, where, for a set P, its *first derived set P'* was defined as the set of limit points of P.[36] Here he expands on the notion in concord with his expanded number system: the derivative $P^{(\gamma)}$ is such that "γ can be any whole number of one of the number-classes (I), (II), (III), etc."[37] That is, γ can be a finite or transfinite number, and thus the set of all of P's derived sets, $P^{(\gamma)}$, includes derived sets of any magnitude. If P is infinite and perfect, then P is non-denumerable,[38] and the set of real numbers as Cantor defined it in 1872 is a perfect set; the collection of derived sets of the real numbers, $\mathbb{R}$, is identical to $\mathbb{R}$ itself.

Thus, with these two conditions, Cantor believed himself to have established a "purely arithmetical concept of a point-continuum,"[39] rather than one based on intuitions or experience, and believed that this notion could then be applied to our understanding of non-mathematical continua such as space or time. Much like the continua of many dimensions discussed in Section 4.3 above, which were analyzed completely in terms of the points contained within them, this definition of continuity relies solely on the connection between the points of the set in question. Thus, one supposes, when analyzing the continuity of time, one would first have to identify the basic elements to be analyzed – perhaps instants – and then we could determine whether the set of instants under consideration was a perfect and connected set. It is of great consequence to note that Cantor's continuity is necessarily one composed of individual elements.

---

[35]See Grattan-Guinness [2001, p. 93].
[36]See Cantor [1872, p. 343].
[37]See Cantor [1883, p. 86].
[38]See Dauben [1990, p. 111].
[39]See Cantor [1883, p. 85].

Cantor's theory of continuity is similar to Dedekind's in some respects. Particularly, Dedekind's set of real numbers can be shown to have the property of perfection. However, Cantor's mathematical concept of connectedness is precisely what Dedekind's theory of continuity lacked, which led to objections that Dedekind had simply presented a collection of distinct objects, and not a continuum at all. While Cantor's connectedness does determine some sort of interlinking between the elements of his continuous sets, his continuum is, like Dedekind's necessarily composed of elements; in order for a set to be judged continuous, one must first have a set of elements to judge, and Cantor's continuity is predicated upon the existence of infinite sets. As we shall see in later chapters, du Bois-Reymond and Peirce were critical of this compositional theory of continuity, and both strove to develop substantially different theories. It is notable that both theories include infinitesimals in crucial roles. Thus, before ending this chapter on Cantor, it behooves us to examine in some detail Cantor's objections to infinitesimal magnitudes.

4.6. Infinitesimals (1883 and 1887)

Cantor was a staunch opponent of the inclusion of infinitesimals into our mathematical systems. In the *Grundlagen*, of 1883, he argued that those who believed infinitesimals to be real quantities were laboring under a confusion; in a letter to Weierstrass in 1887, he formulated the sketch of an argument which was meant to prove a stronger conclusion – that infinitesimals were self-contradictory entities, and thus could not be consistently formulated. In this section, we will examine Cantor's anti-infinitesimal arguments in both of these works.

In the *Grundlagen*, while Cantor vigorously defended the addition of transfinites to our concept of numbers, he took a moment to indicate that this should not imply his acceptance of the existence of infinitesimals. In fact, he believed that those who posited actual infinitesimal magnitudes were confusing the non-genuine infinite with the genuine infinite.

> The infinitely small magnitudes, to my knowledge, have so far been worked out for useful purposes *only* in the form of the non-genuine infinite. [...] On the other hand, all attempts

4.6. INFINITESIMALS (1883 AND 1887)

to transform the infinitely small by force into a *genuinely* infinitely small magnitude should finally be abandoned as purposeless. If genuinely infinitely small magnitudes exist in any other form at all, i.e., are definable, still they surely do not have any immediate connection with the ordinary magnitudes that *become* infinitely small.[40]

Thus, Cantor here claimed that the infinitesimals which appeared in some works of analysis and function theory at the time were not infinitesimals at all, but rather variable magnitudes approaching the infinitely small but at all times actually finite, i.e., examples of non-genuine infinity. Here in the *Grundlagen*, he hedged on whether infinitesimals actually existed, however, writing not that they are impossible entities (as he later would argue), but rather, that if they were somehow definable, they would have no connection to ordinary magnitudes, and thus no connection to the variable magnitudes which are actually useful to analysis.

As the charge of uselessness is a common criticism lodged against theories of infinitesimals, it is worth taking a moment to consider Cantor's stance on the difference between pure and applied mathematics, the criteria for admitting new numbers as actually extant, and the role of use in mathematics – all of which he discusses in his defense of admitting transfinites in to the numerical canon. Cantor's explicit view on usefulness is that pure (or, as Cantor preferred, 'free') mathematics should not be hampered by considerations of whether mathematical creations were useful in the sciences or in any sense whatsoever, and that the most creative and important mathematical advancements were achieved by those who cared not one whit for the usefulness of their theories.[41] While he firmly believed that those working in applied mathematics should indeed consider the metaphysical ramifications of their theories, the criterion restricting the addition of new entities to pure mathematics were looser.

In fact, he wrote that mathematics is "bound only by the self-evident concern that its concepts be both internally without contradiction and stand in

---

[40]Ibid., p. 74.
[41]Ibid., p. 79. Among the mathematicians so identified are Gauss, Cauchy, Dirichlet, Weierstrass, Poincaré, and Bernhard Riemann (1826–1866).

definite relations, organized by means of definitions, to previously formed, already existing and proven concepts."[42] Thus, for the pure mathematician, usefulness is not a necessity when new concepts are introduced; only internal self-consistency, and the relationships the new concept would have with the wider system, need be considered. To meet the latter criterion, Cantor thought that a new concept must be defined well enough to permit distinction from the old concepts.

Cantor argued that transfinite magnitudes meet these criteria, and thus "can and must [be regarded as] existent and real."[43] However, many mathematicians, such as Peirce and du Bois-Reymond have argued that infinitesimals also meet these criteria, and thus, according to Cantor himself, also must be regarded as extant and real. Hence, though Cantor may be correct that genuine infinitesimal magnitudes are "purposeless" and have no "immediate connection with the ordinary magnitudes," according to Cantor himself, neither of these complaints are enough to justify their exclusion from pure mathematics.[44] According to his own criteria, infinitesimals may only be justly excluded if they are proved inconsistent, or if they are incapable of being defined in such a way that they can easily be distinguished from existing mathematical entities.[45]

Cantor possibly saw this flaw in his reasoning, and a few years later he took the more direct approach of attempting to prove infinitesimals internally inconsistent. If he had succeeded, and had shown that infinitesimals could not be consistently defined, then he would have barred them from mathematical reality by the first condition listed above. His argument sketch appeared in a letter to Weierstrass.

---

[42]Ibid., p. 79.

[43]Ibid., p. 79.

[44]In Chapter 7, below, I will argue that Cantor is in fact incorrect about this, and that infinitesimals are mathematically useful.

[45]It must be noted that Cantor believed useless mathematical entities, while they should not be excluded from the canon on the basis of their uselessness, would eventually be abandoned in favor of more fruitful concepts. However, he noted that many concepts had no apparent use when introduced, but soon became indispensable to science; uselessness was, for Cantor, something to be determined over time, not at the moment of mathematical invention. Ibid., p. 79.

"*Linear number magnitudes ζ different from zero (i.e., shortly put, such number magnitudes as can be represented by bounded, straight, continuous segments) which would be less than any ever so small finite number magnitude do not exist, i.e., they contradict the concept of a linear number magnitude.*" The train of thought in my proof is simply the following: I begin from the *supposition* of a linear magnitude ζ which is so small that its product by n, $n \cdot \zeta$, for *every finite whole number n however great* is smaller than unity, and then prove, from the concept of a linear magnitude and with the help of certain propositions from transfinite number theory, that then $\zeta \cdot \nu$ is less than every finite magnitude however small, where ν is an arbitrarily great *transfinite* ordinal number (i.e., *Anzahl* or type of a well-ordered set) from any arbitrarily high number class. But this means that ζ cannot be made *finite by any actually infinite multiplication of any power*, and hence surely cannot be made an *element* of finite magnitudes. But then the supposition made contradicts the concept of a linear magnitude, which is such that every linear magnitude must be thought of as an integrated part of other ones, and in particular of finite ones. Hence, there remains no alternative but to drop the supposition that there is a magnitude ζ which is smaller than $1/n$ for every finite whole number n, and hence our proposition is proved.[46]

The argument has obvious gaps (for example, Cantor does not here specify which "propositions from transfinite number theory" would help us), but the general idea is clear. Infinitesimals, or linear number magnitudes different than zero but less than any finite number magnitude, cannot be made finite by multiplication of any finite *or infinite* number. Therefore, infinitesimals cannot be elements of finite magnitudes – they are off the map, so to speak, and no road whatsoever leads to them. For, to re-quote an essential element of the proof, "every linear magnitude must be thought of as an integrated part of other ones, and in particular, of finite ones."[47] This requirement reminds one

---

[46] See Cantor's 1887 letter to Karl Weierstrass in [Cantor, 1966] with translation by W. D. Hart in [Moore, 2002, p. 306]. The italics are Cantor's.

[47] Ibid., p. 306

of the second condition for allowing new mathematical objects into the canon
– that the relationship to the existing objects be specified.

However, even should a completed version of Cantor's proof work, and establish that infinitesimals were necessarily disconnected from finite quantities, it still would not establish that they are impossible, i.e., intrinsically inconsistent imaginary things. Rather, it would merely prove that infinitesimals cannot be an integrated part of a system of magnitudes, since no finite or infinite multiplication can help an infinitesimal reach the finite numbers of the system. That infinitesimals are isolated from the finite numbers, in that no finite multiplication of an infinitesimal results in a finite number, is a well known feature of infinitesimals, one that lends them their unique character. Transfinite numbers themselves are similarly isolated from the finite numbers: no finite multiplication of a finite number can ever result in a transfinite, and Cantor himself appreciated the isolation of these new numbers.[48]

The key to understanding why Cantor believed infinitesimals were not merely isolated, but self-contradictory, lies in the first sentence of this argument sketch: "Linear number magnitudes ζ different from zero ... which would be less than any ever so small finite number magnitude do not exist, i.e., they contradict the concept of linear number magnitude." It is the notion of an infinitesimal magnitude *qua* magnitude which is, Cantor believed, inconsistent. Infinitesimals cannot be an integrated part of a collection of infinitesimal and finite linear magnitudes, but more importantly, they cannot *consistently* be considered magnitudes at all. In what seems to be an extension of the Archimedean Principle, Cantor defends this supposed inconsistency by claiming that ζ cannot be made finite by any finite or infinite multiplication. And indeed, in the same letter to Weierstrass, Cantor gave a definition of the Archimedean Principle, and insisted that it should not be considered an axiom.

---

[48]See, for example, Cantor [1883, p. 71], where Cantor wrote that "the point at infinity in the complex number plane stands isolated *vis a vis* all finite points." He went on to show that there were infinitely many distinct such points, and further, that it is their isolation from finite quantities which give them the distinctness necessary to justify adding them to the mathematical canon; for were they entirely reducible to finite numbers, the creation of new numbers is wholly unjustified.

4.6. INFINITESIMALS (1883 AND 1887)

The so-called Archimedean Axiom is not an axiom at all but a proposition which follows with logical necessity from the concept of linear magnitude.[49]

As we saw in Chapter 3 above, the inclusion of the Archimedean Principle does conflict with the inclusion of infinitesimals. If the Archimedean Principle follows necessarily from the concept of linear magnitude, then an infinitesimal linear magnitude is indeed a contradiction. Thus, Cantor's argument revolves around one key point; whether non-Archimedean magnitude is possible.

This is not a point Cantor here supports, but rather, simply assumes. And yet, consistent non-Archimedean systems have been developed, both before and after Cantor wrote this letter. Veronese proposed non-Archimedean geometries around 1890,[50] and David Hilbert (1862–1943) proved the consistency of a non-Archimedean geometry and corresponding number system in his *Grundlagen der Geometrie* in 1899. In the twentieth century, Abraham Robinson created a non-Archimedean calculus.

One further note on this proof. Possibly the most controversial point of this argument is that he claims no finite *or infinite* multiplication can integrate infinitesimals with finite magnitudes, for soon after Cantor's *Grundlagen*, in which he introduced transfinite magnitudes, mathematicians were defining infinitesimals as the multiplicative inverses of transfinites.[51] Abraham Robinson's non-standard analysis does something similar, and incorporates both infinitesimal and infinitely large quantities into the same system with finite quantities. Systems of magnitudes which include infinitesimals – such as the systems of Paul du Bois-Reymond and of Charles Sanders Peirce – are decidedly non-Archimedean. Thus, as we will soon see, these systems require a theory of continuity and even a theory of real numbers substantially different from those of Cantor and Dedekind.

---

[49]See Moore [2002, p. 306].
[50]See the biography of Veronese at the University of St. Andrews, [Online]www-groups.dcs.st-and.ac.uk/history/Biographies/Veronese.html.
[51]Cantor, 1883.

CHAPTER 5

Paul du Bois-Reymond

5.1. Biography and Introduction

Paul du Bois-Reymond was born in 1831 in Berlin. His father was from Neuchatêl, an area of modern-day Switzerland which was at the time nominally part of Prussia. The du Bois-Reymonds spoke French at home. Emil, Paul's elder brother, was a noted physiologist. Paul followed his brother into the physical sciences and medicine, studying at the Collège in Neuchatêl, the Gymnasium in Naumburg, and the University of Zurich. While studying in Könisberg, he changed his focus to add more of a mathematical element, and wrote his doctoral thesis at the University of Berlin in mathematical physics, focusing particularly on mathematical analyses of the movements of liquids.[1]

Paul du Bois-Reymond's physical sciences background brings an element of practicality to his mathematics and his philosophy; many of his examples are drawn from the physical world. Applied mathematics was the focus of much of his career; considerations of how theories match up to their applications dominate much of his philosophical work. His applied and even physical approach to mathematics is an important counter-balance to the more theory-based work of the other three mathematicians being considered in the present work. Interestingly, though his mathematical and philosophical theory is well grounded in application, many of his mathematical and philosophical contributions center on things far removed from the physical world: infinite series, infinite functions, and transfinite and infinitesimal quantities, things he himself admits are not grounded in or reducible to physical experience.

---

[1] See du Bois-Reymond [1859].

It was his study of applied mathematics that led to his interest in partial differential equations and functions with infinite domains. He was the first person to give an example of a continuous function of a Fourier series which diverged at every point. Much of his career was spent creating his Infinitärcalcül, or Infinitary Calculus, a theory whereby functions with infinite domains and ranges were compared and ordered. In particular, the ratio of two functions, both with infinite domains, was subjected to the limit operation, which produced interesting results; we shall examine this Infinitärcalcül further on in the chapter as it has philosophical ramifications related to continuity.[2]

The work of du Bois-Reymond with which we are most directly concerned is the 1882 work, *Die allgemeine Functiontheorie*. Written late in his life (he died in 1889), it is a philosophical attempt to show that mathematics is a science like any other. As he writes in the introduction of *Die allgemeine Functiontheorie*, fundamental mathematical concepts are not as rigorously established as key concepts in other sciences, instead left to be intuited. Problematically, our mathematical intuitions often conflict: "What mathematician could deny that ... the concept of limit and its near-parents, those of the unlimited, the infinitely large, the infinitely small, the irrationals, etc, still lack solidity!"[3] He argued that teachers and researchers alike gloss over these concepts rather than defining them rigorously, and then make use of them freely in calculus as though they had been proven. Du Bois-Reymond wished to solidify the basis upon which our mathematics is built by closely examining the intuitions behind all of these concepts.

In this chapter, we will briefly consider the details of the Infinitärcalcül and suggest how this system inspired du Bois-Reymond to consider the foundations of mathematics an important field of inquiry, then discuss in detail du Bois-Reymond's attempt in *Die allgemeine Functiontheorie* to define rigorously

---

[2]For more information on Paul du Bois-Reymond's life, see the University of St. Andrews biography, [Online]www-groups.dcs.st-and.ac.uk/history/Biographies/Du_Bois-Reymond.html.

[3]See du Bois-Reymond [1887, p. 21]. The original text of this book was published in 1882, and was titled *Die allgemeine Functiontheorie*. In this chapter, references will be to the French translation. The English is the author's translation from the French unless otherwise noted. In the introduction to the French translation, du Bois-Reymond mentioned that he had the opportunity not only to oversee the translation and ensure its accuracy, but that he made several changes to the content of the work itself; thus the preference for the translation over the original.

the concept of limit and its near-parents, continuity, infinity, and infinitesimals. Du Bois-Reymond believed that at bottom, our mathematical concepts were founded on two competing intuitions, an Idealist and an Empiricist intuition, and the first third of the book represents the argument between these two camps. We will spend some time with the Idealist/Empiricist argument, as it gives us great insight into what du Bois-Reymond saw as the fundamental philosophical problems behind the basic concepts of mathematics. A final section will draw together du Bois-Reymond's own conclusions about these basic concepts, discuss how these concepts interact in his theory of continuity, and briefly compare his theory to those of Cantor and Dedekind.

5.2. Infinitärcalcül

Du Bois-Reymond's Infinitärcalcül was in essence an attempt to order functions with infinite domains and ranges using the limit theorem. He began this project in an 1870-1 paper,[4] and developed it through many papers written over many years. The mathematical basics of the Infinitärcalcül are as follows.[5] First, let f and g be functions with infinite domains and ranges; we can then take the limit of their ratio, thus:

$$\lim_{x \to \infty} \frac{f(x)}{g(x)}$$

The results of computing this limit tell us something about how the two functions relate to each other. If the limit approaches infinity, then we say that $f(x) \succ g(x)$. If the limit approaches zero, then we say $f(x) \prec g(x)$. If the ratio has a finite, non-zero limit then we say $f(x) \approx g(x)$. We could say that, if it were absolutely clear we were only speaking in analogies, in the first case, $f(x)$ is infinitely bigger than $g(x)$, in the second case $g(x)$ is infinitely bigger, and in the third case they are equivalent. Du Bois-Reymond viewed this result as producing a rough ordering of such functions, and attempted to prove a more rigorous ordering and to draw results from this organization of functions. As he

---

[4]See du Bois-Reymond [1870, p. 338–353].

[5]This summary is drawn from Gordon Fisher's article [Fisher, 1981] and G. H. Hardy's book [Hardy, 1924].

wrote in his first paper on the subject, "This new algorithm which shows some analogy with ordinary inequalities, can be called one of *infinitary inequalities*."[6]

Thus, for example, take the function $f(x) = x$, and the function $f(x) = x^2$ (or, to abbreviate, the functions x and x^2). "The student soon learns that as x tends to infinity ($x \to \infty$), then also $x^2 \to \infty$, and moreover, that x^2 tends to infinity *more rapidly than* x."[7] The magnitudes of infinity[8] that characterize du Bois-Reymond's infinitary inequalities are just these: that some increasing functions approach infinity more rapidly than others, and that some decreasing functions approach zero more rapidly than other decreasing functions. Thus, applying our above definition of the $\succ$ relationship, the limit of x/x^2 is zero, and thus, $x \prec x^2$. The inverse relationship also holds in this case; as x grows, x^2/x approaches infinity, and thus $x^2 \succ x$.

The above example provides us with a way of generating many such magnitudes of infinity, in fact infinitely many. Consider not only x and x^2, but also x^3, x^4, etc., insuring as the exponents increase that the function approaches infinity even more rapidly than the previous function. Noting that there are thus infinitely many magnitudes of functions, du Bois-Reymond thus wished to define a series of infinitary inequalities

$$f_1(x) \prec f_2(x) \prec f_3(x) \prec \cdots$$

Du Bois-Reymond envisioned this infinite ordering of functions as closely analogous with the real numbers, and wished to prove the density of the ordering to further the resemblance:

> Just as between two functions as close with respect to their infinities as one may want, one can imagine an infinity of others forming a kind of passage from the first function to the second,

---

[6]See du Bois-Reymond [1870, p. 339], as quoted in Fisher. At the time, no algorithm existed, but only notation. Fisher suggests du Bois-Reymond might have been anticipating algorithms he intended to develop. See Fisher [1981, p. 102].

[7]See Hardy [1924, p. 1].

[8]These magnitudes of infinity, note, have nothing to do with *transfinite* numbers, unlike Cantor's differing magnitudes of infinity; these functions all range over the real numbers and thus have the same cardinality as the reals themselves. The Infinitärcalcül thus provides an ordering of infinities of the same cardinality.

one can compare the sequence F [a scale of infinity] to the sequence of real numbers, in which one can also pass from one number to a number very little different from it by an infinity of other ones.[9]

Thus, in his 1870–1 article, du Bois-Reymond established a family of functions, invented a system of ordering them, and made some indications on how density could be approximated. In subsequent articles, he drew consequences from this ordering and developed theorems for his calculus of functions. By 1875, he began explicitly to address philosophical considerations surrounding the Infinitärcalcül. In the 1875 article, he finally committed himself to accepting the concept of the actual infinite, and began to speculate on the relationships between different types of functions (increasing versus decreasing, for example), as well as on the nature of functions and the concept of limit – essential tools in his theory of infinitary functions[10]

An 1877 paper, "Ueber die Paradoxen des Infinitärcalcüls," was dedicated to investigating the similarities and differences between this arranged chain of functions and the real numbers. Within this article, he claims that the continuity of his Infinitärcalcül is comparable to the continuity of the real numbers, which gives us further insight into the Infinitärcalcül itself, but more importantly for our current purposes, gives us insight into his early ideas of mathematical continuity and the creation of the real numbers.

> Thus through more precise consideration the rational numbers always approach more closely to one another, yet in our minds gaps are always left between them, which mathematical speculation then fills with the irrationals.[11]

Notably, in general outline this account is similar to Dedekind's definition of real numbers and mathematical continuity, but the overtones are quite different. The density of the rationals is reached through "precise consideration," while the gaps between them are not mathematically proven, as in Dedekind,

---

[9]Ibid., p. 343.
[10]See Fisher [1981, p. 105].
[11]See du Bois-Reymond [1877, p. 150], as quoted in [Fisher, 1981, p. 107]; presumably, the translation from the German is also Fisher's.

but rather the gaps occur "in our minds." The real numbers do not arise as a mathematical necessity as in Dedekind, but rather through "mathematical speculation," clearly implying a lack of rigor and justifiability, which is to be contrasted with the precision of the rationals.

Indeed, du Bois-Reymond went on to claim that irrational numbers were introduced only to complete the comparison between numerical sequences and geometric ones, and depend intimately on the limit concept, which itself needed further analysis and investigation.[12] As we shall soon see, this is precisely where the *General Theory of Functions* begins: with an effort to comprehend these mathematical concepts at the most basic level. This work is du Bois-Reymond's attempt to philosophically ground the continuity of the real numbers, the concept of limit, the relationships between geometry and systems of numbers, and of course, the nature of functions themselves.

5.3. Idealist versus Empiricist: Basic Theories, and Straight Lines

Du Bois-Reymond began his *Die allgemeine Functiontheorie* by considering two philosophical approaches to mathematics; that of the Idealist, and that of the Empiricist. These terms do not refer to theories of actual philosophers or mathematicians; rather, du Bois-Reymond believed that each of us who thinks carefully about mathematics will find both empirical and idealistic tendencies within ourselves, and that in fact, this split in intuitions is the very reason we have such trouble elucidating the most basic elements of our mathematics.[13] As he wrote in his introduction, "There is, in the mind, two completely distinct manners of apprehending things, which have an equal right to be taken for the fundamental intuition of exact science."[14]

---

[12] See Fisher [1981, p. 108].

[13] Though du Bois-Reymond clearly treats both viewpoints as innate to us all, and two sides of the same mathematical intuition coin, he personifies each of them as though they were separate people, each holding to only their own basic assumptions, and stages much of the book as a dialogue between them. We shall follow suit in the rest of this chapter and refer to "the Idealist" and "the Empiricist" as though they were individuals with distinct and often contradictory theories of mathematics, rather than as the internal conflict he believed most of us felt on the subject.

[14] See du Bois-Reymond [1887, p. 22].

5.3. IDEALIST VERSUS EMPIRICIST: BASIC THEORIES, AND STRAIGHT LINES

Much of this book is spent drawing out the logical consequences of the Idealist versus the Empiricist worldview (if we may call it that) as concerns our mathematical tendencies. Du Bois-Reymond thus considers the mathematical status of infinitesimals, the continuity of the geometrical line, and the relationship between the geometrical line and the number line, from within each framework, beginning with the original intuition for each side. As we will see, the conclusions we reach on all of these topics are radically different, depending on which framework we adopt.[15] Thus, we shall look at each philosophical viewpoint in some detail in subsection 5.3.1. We will next apply these viewpoints to a particular case, that of the straight line, for three reasons: first, this particular example clearly demonstrates how different the two positions are; second, the different theories of the straight line are relevant to the differing conclusions about the plausibility of infinitesimal quantities; and third, our discussion of du Bois-Reymond's theory of continuity shall also depend largely on these intuitions about the line itself.

5.3.1. The Basic Viewpoints and Their Fundamental Theses. In basic outline, the Idealist represents our belief in logic, idealized geometrical figures, and precision that is beyond that which we can accomplish physically. The Empiricist represents our belief in the fundamental connection between mathematics and the world of sensation, of things that can be seen and felt and physically manipulated. Thus,

> Idealism believes that the truth of certain limited forms of our ideas is required by our understanding, though they may lie outside of all perception and sensory representations. Empiricism is the system of complete abnegation; it admits only as extant that which can be perceived or reduced to perception.[16]

Empiricism thus stems from our intuitions that the point of mathematics is to quantify and explain our world, and begins with what is directly experienced, or at least to forms which can be reduced to such experience. This recalls

---

[15]It is my belief that du Bois-Reymond favored the Idealist intuitions, if one must choose, but that he strove to create a synthesis of the two that would retain the power and creative possibilities of the Idealist view while remaining at least somewhat grounded in Empiricist principles.

[16]See du Bois-Reymond [1887, p. 22].

Dedekind's assertion that mathematics begins when we begin to count, assuming that we are indeed counting real world objects, at least at the beginning. The Idealist, on the other hand, privileges our wishes to extend mathematics as far as logic and our imagination can take us, and to create mathematical objects which satisfy our logical intuitions, whether or not these objects have any basis in reality.

To illustrate the difference, for the Empiricist, geometry can contain shapes such as circles because we experience round things in our daily life. This is reminiscent of Plato's definition of shape in the *Meno*; "A shape is that which limits a solid; in a word, a shape is the limit of a solid."[17] Plato's definition could apply to geometrical, idealized solids, or it could apply to actual solid objects in the world; the shape of an object is the limit of that particular object, for du Bois-Reymond's Empiricist, and a shape in general, such as a circle or a triangle, is a mental collection of all that is similar in the shapes of circular or triangular objects experienced. We must, however, be careful how we abstract away from our experience; the Empiricist believes the Idealist abstracts so far away from experience that it is impossible to return to it.

For the Idealist, on the other hand, the allowable mathematical objects are those that meet certain stringent logical criteria, whether or not they are connected in any way whatsoever to our experience. The Idealist's geometry is thus quite Euclidian, with partless points, breadthless lengths, and the Idealist would most likely agree with Euclid's assertion that "a circle is a plane figure contained by one line such that all the straight lines falling upon it from one point among those lying within the figure are equal to one another, and the point is called the centre of the circle."[18] Thus, a circle, for an Idealist, is not a generalization of the shape of all circular objects, abstracted away from particulars such as color and location, and perhaps also extracted away from flaws, but it is rather an idealized object, constructed from already idealized objects such as point, line, and plane. This Idealist circle is not based on experience, nor is it reducible to experience; and further, it cannot physically exist at all. Any particular representation of a circle fails to meet the stringent

---

[17] See Plato [2002, p. 65 (76a)].

[18] See Euclid [1956, p. 153–154]. In fact, the Idealist defines a sphere in a similar way to this. See du Bois-Reymond [du Bois-Reymond, 1877, p. 95].

5.3. IDEALIST VERSUS EMPIRICIST: BASIC THEORIES, AND STRAIGHT LINES

logical criteria of this definition: any drawn circle is drawn of a curved line with width, is likely to fail to be precisely round, etc.

5.3.2. Two Theories of the Straight Line. An important difference in Idealist and Empiricist mathematics is how they each view the nature of the straight line itself. For the Empiricist, the line is one drawn from experience; the idealized, one-dimensional, and infinitely long straight line does not exist in nature, and is not deduced from or reducible to representations, therefore it does not exist.[19] In Idealist thought, a straight line is an idealized geometrical object, infinite in extent, infinitely divisible, existing only in one dimension.[20] The ramifications for each viewpoint are worth going into in detail, as Du Bois-Reymond argued that the Idealist theory of the straight line leads directly to the logical necessity of infinitesimals, as we shall see in the next section. The Empiricist theory, on the other hand, leads to a conception of the straight line which is quite unique.

The Idealist begins with this geometrical idealization of a line to draw a correspondence between the real numbers and the points on this line. He does this as he recognizes that precise measurement of geometrical objects is "fundamental to the science of magnitude."[21] Points, it must be noted, are themselves idealized objects; as they lack length, they therefore have no extension at all. For simplicity, the Idealist considers only the line segment between the points 0 and 1 – the "unit length."

The "rational lengths" – points on the line which correspond to rational numbers, and which mark out lengths which are a particular percentage of the unit line – are easy to find. Thus, we can easily locate "halves, thirds, quarters, etc. ... and also multiples of these fractions of the unit length."[22] We can also easily construct innumerable irrational lengths, such as the roots of rational

---

[19]See du Bois-Reymond [1887, p. 90].

[20]The Idealist specifies two different ways of conceiving of the idealized line and point. One is to begin with an idealized plane, and define the line as an intersection of two planes, and the point as the limit point of a line segment. The other is to begin by defining the point as space which is infinitely contracted, such that it has no extension in any direction, and then creating a line segment by following the motion of this point through space in one direction. See du Bois-Reymond [1887, p. 96–97].

[21]Ibid., p. 64.

[22]Ibid., p. 64.

numbers and multiples of these roots, and "the unit length is covered in our representation by a more and more dense collection of points."[23] However, this process will never completely fill the unit length with points.

> On the contrary, two neighboring points always remain separated by a segment of straight line which, abstracting away from its length, completely resembles the unit length. This is the image which always accompanies my representation of magnitude.[24]

This Idealist representation of magnitude is thus that of a line segment populated by infinitely many points which each correspond to a real number, and which are such that each and every pair of points maintains a line segment between them. Further, no matter how many points populate a line, it is never *composed* of them. "I reject the enlargement of the concept of magnitude according to which the line must be composed of points, the surface of lines, etc."[25]

This representation of a line segment is common enough, but there are at least two interesting features which should be noted. First, the only irrational numbers explicitly corresponding to points were roots of rational numbers and their multiples, as the Idealist believes, as does du Bois-Reymond himself, that most if not all irrational numbers are themselves limits;[26] the Idealist is not guaranteed that points corresponding to any mathematical limit actually exist.[27] Second, mere density is not the essential feature of the Idealist's magnitude; rather, it is the existence of a line segment between any two points that is precisely identical to every other line segment, except in regards to their size. The importance of this feature leads the Idealist also to accept the existence of infinitesimals, as we shall later see.

---

[23]Ibid., p. 64.

[24]Ibid., p. 64.

[25]Ibid., p. 70.

[26]This is a long-held belief on du Bois-Reymond's part. See, for example, du Bois-Reymond [1877, p. 152] and Fisher [1981, p. 108].

[27]He will eventually prove the existence of these point-limits, but not until he has proven the existence of infinitesimal quantities; thus, infinitesimals are necessary for the geometrical line and our number system to be truly comparable.

5.3. IDEALIST VERSUS EMPIRICIST: BASIC THEORIES, AND STRAIGHT LINES 93

First, though, let us look carefully at the straight line of the Empiricist. As we saw in 5.3.1, du Bois-Reymond characterized the Empiricist viewpoint as one of "complete abnegation,"[28] thus characterizing it mainly by what it denies. To be precise, it denies the existence of any mathematical object not connected to or reducible to our experience. In fact, the title of the section in which the Empiricist begins to enunciate his own view is, "Purification of the System of Concepts."[29] However, for a system of complete abnegation, the Empiricist's ontology is rather rich, and many positive claims are made.

For example, after denying the existence of idealized geometric objects,[30] the Empiricist commits himself to the existence of non-idealized geometric objects. Thus, points are not idealized bits of non-extended space; they have extension. Lines are not idealized one-dimensional objects that exist only in the mind, but all lines have a thickness as well as a length. In fact, the thickness of the line is related to the extension of the point; the line can be as thin as you wish (as long as it is not infinitely thin, which makes no sense to the Empiricist), but as a point is "the portion of space common to two lines which intersect,"[31] the point thus has precisely the width of the lines, and it is just as tall as it is thick. In fact, as it is possible for three lines to intersect in three-dimensional space, the point is also as wide as it is tall and thick; it forms a very small sphere. Consequently, there is no contradiction viewing the line as composed of points. Recall Aristotle's argument against the compositionality of the line from Chapter 2: the main engine of the argument was the assumption that points have no extension, no parts, and therefore it is impossible for them to be next to each other. The Empiricist point has no such restrictions. Therefore, he concludes, the line "is composed of points without gaps."[32]

It is important to emphasize that the line (and therefore the point) does not have a specified thickness, but rather, it can be as thin *as one wishes*.[33] Traveling toward infinity is not a problem for the Empiricist; only reaching it is a problem. Thus, while the Idealist admits any actually infinite set which has a

---

[28]See du Bois-Reymond [1887, p. 22].

[29]Ibid., p. 102.

[30]"It is absolutely forbidden to deduce whimsically from our representations the idea of the perfect straight line." Ibid., p. 105.

[31]Ibid., p. 105.

[32]Ibid., p. 106.

[33]In the French, *à volonté*. See, for example, du Bois-Reymond [1887, p. 105, 106].

logical rule describing it, the Empiricist only allows that, using the rule, we can continue to generate as many numbers *as we wish*. Otherwise, we are at risk of postulating that the numbers, once we give them a law, proceed "separated from human mind to continue all alone their route toward the infinite," or that the rule itself is identical to an actually infinite set, both absurdities.[34] Thus, instead of infinite, the Empiricist prefers 'as large as one wishes;' instead of infinitesimal the Empiricist would substitute 'as small as one wishes;' and instead of idealized geometric shape, the Empiricist would use 'as perfect as one wishes.'[35]

5.4. Infinitesimals, For and Against

The Idealist, according to du Bois-Reymond, is logically committed to the existence of infinitesimals, and so, in addition to arguing for their existence, the Idealist provides us with an analysis of their properties and how they fit into the larger mathematical universe. This section shall have three subsections: an analysis of the Idealist's argument for the existence of infinitesimals, a discussion of the properties of these infinitesimals, and the Empiricist's counter-argument.

5.4.1. The argument for infinitesimals. The argument turns on the Idealist's theory of magnitude already discussed above. I shall quote the argument in its entirety before looking at each step in turn.[36]

---

[34]Ibid., p. 84.

[35]This is reminiscent of Aristotle's view that the actual infinite did not exist, but that mathematicians need not despair as they "they do not need the infinite and do not use it." Aristotle, *Physics*, Book III, 207b p. 30.

[36]The original German of this proof is as follows:

Denn halten wir fest, was wir oben als korrekten Grössenbegriff hinstellten, daß Punkte auf der Länge nicht ohne Abstand aufeinander folgen, also nicht aneinander stoßen können, sondern immer durch Strecken getrennt sind, daß also bloße Punkte nie eine Strecke bilden können, so sind auch die unendlich vielen Punkte durch unendlich viele Strecken getrennt, und von diesen Strecken kann keine endlich, das ist in endlicher Zahl in der Einheitsstrecke enthalten sein, weil bei der Willkürlichkeit der Längeneinheit jede noch so kleine Strecke die nämliche Beschaffenheit, wie die Längeneinheit haben muß, so daß auf ihr wieder unendlich viel Teilpunkte vorhanden sein müßten.

Es ergiebt sich also, daß die Einheitsstrecke in unendlich viele Teilstrecken zerfällt, von denen keine endlich ist. Also existirt das Unendlichkleine wirklich. (p. 71–72 of the German edition).

> The proposition that the number of points of division of the unit length is infinitely large produces with logical necessity the belief in the *infinitely small*.
>
> In fact, we have established above that, according to the true concept of magnitude, these points do not follow each other without an interval, thus they cannot be united but are always separated by extensions, so that points alone can never form extensions; therefore infinitely many points are separated by infinitely many extensions. Thus, of these extensions none can be finite, which is to say cannot be contained a finite number of times in the unit length, because the unit of length being arbitrary, every extension as small as it may be must be organized like the unit length, and similarly contain infinitely many points of division.
>
> One thus sees that *the unit extension is decomposed into an infinity of partial extensions, of which none is finite. Thus the infinitely small really exists.*[37]

The heart of this argument is deceptively simple: if you take an object that is finite in extension, and divide it actually infinitely many times, the resulting pieces must be infinitely small. In greater detail: recall, according to the Idealist, points do not compose the line. They can be viewed as locations or points of division upon the line; dividing the unit length into equal halves, or dividing it precisely at the three-quarter mark or the square root of 2, or marking out $1/35^{\text{th}}$ of the line. These points "cannot be united," but rather, between any two points, no matter how close, an extension must always remain. Given that there are actually infinitely many such points of division upon the unit length, we may divide by all of them simultaneously; we may locate all of them, on the line, at once.[38] As there must be a line segment between each of these infinitely many points, there must be infinitely many line segments, and each of them must be infinitely small. The argument, then, may be sketched as follows:

---

[37] See du Bois-Reymond [1887, p. 73].

[38] Notice that this is a physical impossibility; we cannot physically complete an infinite task, nor may we draw infinitely many points on a line; but while the Empiricist would object to this, the Idealist does not require us to complete this task. The first task itself "take a unit segment of the line" is itself physically impossible to do, as such one-dimensional objects are ideal and exist only in the mind's eye. The Idealist is only concerned with the logical possibility of infinite division, not the physical possibility.

i) There are infinitely many points of division on a unit segment.
ii) Between any two points exists a segment.
iii) Therefore, division by all of the points at once produces infinitely many infinitesimal segments.

The Empiricist, of course, would object to this argument at all three stages, denying the existence of actually infinitely many points upon the line, as well as infinite division, as these are not concepts which can be reduced to empirical experience. Further, it is possible for an Empiricist to view a line as solely composed of points, with each point having a physical presence and touching the next one. Thus, the Empiricist is not committed to there always being a segment between any two points, and without premise 2 above, division by all points at once would result in simple decomposition into the points themselves, rather than leaving behind these infinitesimally small bits of detritus. Finally, the Empiricist would object to the acceptance of infinitesimal segments regardless of the argument which seemed to produce them, as the Empiricist does not believe that objects resulting from logical manipulation of idealized objects are metaphysically acceptable; we may only allow objects stemming from or reducing to empirical experience into our ontology, which infinitesimals clearly fail.[39]

The argument is not only objectionable to the rather extreme Empiricist viewpoint, it also requires philosophical commitments which not all modern mathematicians would accept, and which Cantor and Dedekind would both positively reject. They would accept the first premise – an actual infinity of points[40] – and they would accept that there is a segment between any two points on the line. However, as proponents of a compositional line,[41] they would not

---

[39]It is worth noting that if du Bois-Reymond is correct and we all have Empiricist, as well as Idealist, intuitions toward mathematics, that may account for some of the deep uneasiness we feel when contemplating these rather unusual mathematical objects. For all of their purported usefulness in a variety of situations, we simply have no experience, nor can we have any experience, of magnitudes smaller than any finite magnitude, that are nevertheless greater than zero.

[40]Interestingly, as we shall see in the next chapter, in Charles Sanders Peirce's final and most fully developed theory of continuity, he would not accept this premise. He argued that a true continuum contained no points whatsoever, and reached this conclusion through reasoning very similar to that which du Bois-Reymond uses to establish his second premise.

[41]Cantor more explicitly than Dedekind, with his "point continuum," but as Dedekind assumed an equivalence between the number line and the geometrical line, and the former

5.4. INFINITESIMALS, FOR AND AGAINST

agree that the conclusion follows, and their reasoning may follow the reasoning given by the Empiricist, above. As their segment is itself composed entirely of points, a line is nothing but points, no extra-metaphysical segments remain outside of the set of the points themselves. Thus, were it possible to divide the unit segment by all of its points at once, one would be left only with the collection of points, with nary a segment between.

We shall address du Bois-Reymond's theory of continuity explicitly and in greater detail below, but some unpacking here is necessary to fully understand the meaning behind his second premise, to ensure his conclusion follows. Notice that in the argument itself, he draws explicitly on the "true concept of magnitude." Let us assume that this "true concept of magnitude" refers to the Idealist's "image which always accompanies my representation of magnitude"[42] discussed in the previous section. Thus, recall, the essence of linear magnitude for the Idealist was that any two "neighboring" points were separated by a segment, which itself was identical to the unit segment in every respect except for size. Recall from 5.3.2 (especially note 20) above that the Idealist not only explicitly rejects the composition of the line from points, but constructs the line either from an intersection of two planes, or the motion of a single point as it travels. The line, in its most basic essence, must have length. Adding to this the notion that each part of the line must itself resemble every other in every respect except for size, it follows that every part of the line must itself have length.[43]

Thus, a point is not a part of the line, the line and its parts must all have something the point does not – extension. We can divide a line as many times as we wish, but attempts to decompose the line, to deconstruct it, even by simultaneous infinite division of all of its points, cannot destroy this essential feature; the resulting parts must themselves have extension. Simultaneous infinite division only ensures that these parts are themselves infinitely small; extension remains.

---

was composed entirely of numbers, one assumes he believed the latter was composed entirely of points.

[42]Ibid., p. 64.

[43]As we will see in the next chapter, Peirce as well believed this "mirror" feature, every part resembling every other part in every respect except for size, was necessary for continuity. Peirce attributes his inspiration for this feature of continuity to Immanuel Kant.

As a side note, even with a compositional construction of the line from points, some version of du Bois-Reymond's argument must, it seems, follow from his two premises if we allow infinite division. For even if the line is composed of points, and line segments are themselves simply subsets of the points of the entire line, one can divide the unit segment by an actual infinite number of points without decomposing it into its basic elements. If we assume the Dedekind/Cantor postulate is correct, and that there is a point on the geometrical line for every real number, and vice versa, then the points on the unit segment correspond exactly to the numbers between zero and one. Suppose, then, that we divided the unit segment by all of the points corresponding to rational numbers. This would effect an actual infinite division without decomposing the line into its component parts; all of the points corresponding to the irrational numbers would remain. The segments thus produced by such an infinite division of a finite extension would have to be infinite in number, and infinitely small in length.

5.4.2. Some Features of Infinitesimals. Briefly, I wish to overview some of the features of the Idealist's infinitesimals, so that we may better understand the nature of these infinitely small segments and their place in mathematics.

"*Infinitesimals are non-zero.*"[44] This follows immediately from the assumptions we have already seen, for if the infinitesimal line segments had a length of zero, the Idealist would be left with a line composed merely of points. For zero is not a magnitude, but rather, it represents the lack of all magnitude, whereas infinitesimal segments have magnitude.

"*A finite number of infinitesimal extensions added to each other do not equal a finite extension, but an infinitesimal extension.*"[45] Du Bois-Reymond calls this the "principle property of the infinitely small,"[46] and it implies either that there are different sizes of infinitesimals, or that the addition of an infinitesimal to an infinitesimal does not lead to an increase in magnitude. Thus, assume γ and δ are infinitesimals. According to this principle, either $\gamma + \delta = \gamma$, or $\gamma + \delta$ equals a third infinitesimal. If the first is the case, then it is also likely

---

[44]Ibid., p. 78–79.
[45]Ibid., p. 73.
[46]Ibid., p. 73.

5.4. INFINITESIMALS, FOR AND AGAINST

that $\gamma = \delta$, i.e., that there is only one infinitesimal. There is some indication that the previous interpretation holds, as in the discussion of this property, du Bois-Reymond indicates that it seems to violate our concept of equality, and thus he limits the concept of equality to finite quantities: "I call equal two finite extensions when there does not exist between them any finite difference."[47] The Idealist does not here provide us with conclusive evidence for one interpretation over another. Either way, one key inference to be drawn from this property is that finitely many infinitesimals (or finite reiterations of the same infinitesimal) will never reach a finite magnitude, just as finitely many finite magnitudes will never reach beyond finitude.

"*Two finite quantities differing by only an infinitesimal are equal.*"[48] This is related to the above definition of equality, but it is also closely related to the next property, that "*a finite quantity does not change if one adds to it or subtracts from it an infinitesimal quantity.*"[49] This is a common property of infinitesimals; that adding or subtracting them to or from finite quantities does not increase or lessen the finite quantities. Thus, $1 + \gamma = 1$. These two properties are particularly noteworthy here as they explicitly introduce the notion of infinitesimal *quantities*, whereas before the Idealist limited his discussion to infinitesimal *magnitudes* or extensions.[50] Thus, one can assume that in addition to infinitesimally small segments, du Bois-Reymond's Idealist is logically committed as well to infinitely small numbers. One can imagine these numbers as the measurement of these segments.

"*The infinitely small is a mathematical quantity and has in common with the finite the set of its properties.*"[51] This is the most intriguing of the Idealist's properties of infinitesimals. First, the Idealist now explicitly defines infinitesimals as mathematical quantities, rather than geometrical extensions. This is consistent with the Idealist's view of number as measurement; if there is a geometrical magnitude to measure, there is a corresponding mathematical quantity which measures it. (Recall, it was the other relationship which did not necessarily hold – a limit of a mathematical series, such as an infinite decimal expansion, did not necessarily correspond to a limit-point on the straight

---

[47]Ibid., p. 74.
[48]Ibid., p. 74.
[49]Ibid., p. 75.
[50]Ibid., p. 74.
[51]Ibid., p. 75.

line). Thus, it is notable, but not astonishing, that du Bois-Reymond's Idealist has now explicitly introduced infinitesimal quantities as well as infinitesimal magnitudes.

It is, on the other hand, astonishing to claim that the infinitesimal quantity "has in common with the finite the set of its properties," especially after we have enumerated the properties infinitesimal magnitudes have which finite magnitudes do not: that two finite quantities differing only by an infinitesimal are equal, and that a finite quantity cannot be changed by the addition or subtraction of an infinitesimal quantity. At the very least, infinitesimals do not behave as finite numbers when they interact with finite numbers.

However, one can safely assume that du Bois-Reymond's Idealist does not mean that infinitesimal quantities have in common with finite quantities every single property. The properties discussed above, in particular, are properties which arise when quantities of different types interact – infinitesimal versus finite. Similar properties come to light when comparing finite quantities and transfinite ones. Du Bois-Reymond stated, in the discussion of this last property of infinitesimals, that the science of comparing different types of quantities to each other is called the "*calcul infinitaire*"[52] – the Infinitärcalcül discussed in Section 5.2 above. He notes that the addition of infinitesimal quantities to our mathematics opens the door for this comparison between types of quantities. In fact, the Idealist believes that transfinite quantities can be proved from the existence of infinitesimal ones (they are simply the inverse of each other).

Given that the peculiar properties of infinitesimal quantities arose during calculations with finite quantities, one can assume that by this last property of infinitesimals, the Idealist means for infinitesimals to have properties *within their type* similar to the properties of finite quantities. This feature may, therefore, solve our quandary above about whether there were different magnitudes of infinitesimals, or if all infinitesimals were equal to each other. One key property of finite quantities is that if we add them to each other, they produce a new finite quantity: $3 + 4 = 7$. Thus, it seems at least sensible to assume that du Bois-Reymond meant for there to be more than one infinitesimal quantity, if they are to mirror the finite in its key properties, and thus the addition of two infinitesimals should produce a new, and larger, infinitesimal.

---

[52]Ibid., p. 75.

There is one last feature of infinitesimals to consider before we move on, though du Bois-Reymond himself does not address this. If indeed every segment of the line is to mirror every other, in every feature other than size, it seems as though the infinitesimal segments produced by his infinite division should do so as well. A highly important feature of a segment of a line, indeed the very feature on which this argument is based, is that it is infinitely divisible. It should therefore follow that an infinitesimal segment is itself infinitely divisible. If this is the case, then this would be one more way in which the set of infinitesimals mirrors the set of finite numbers, and would imply that there are many different magnitudes of infinitesimals indeed – in fact, possibly, nondenumerably many.[53]

5.4.3. The Empiricist's Response. As we saw briefly above, the Empiricist does not accept infinitesimals at all; they are not drawn from experience, they are not reducible to experientially-derived principles. As the Empiricist states, "I believe that we are not authorized to admit to or to create in our rational mathematics things of which we neither have nor could ever have representations."[54] Furthermore, he does not accept the premises of the Idealist's argument. He does not agree that "points alone can never form an extension," nor that there are infinitely many division points on a line. There are as many division points on a line as we need, and no more; it makes no sense to extend the division of a segment beyond practicality. We are told, by the Empiricist:

> We could represent equal parts [of the unit length] in innumerable quantity, each as small as one wishes, precisely because inexactitude diminishes as we are permitted to push the division of an extension as far as we wish. But the pursuit of this division, when we do not halt at any size, finishes by being lost in vague hand waving. Thought can push this division as far as we want, but this act does not prove the infinitely small, but

---

[53]Note, however, that the result of the infinite division is not actual line segments in their full glory, as this would be impossible, but partial segments. They can have a beginning point, but necessarily, no end point. Given that they do indeed have extension, however, I cannot help but think that they also must be divisible, though the implications of dividing a segment by all of its points only to gain as a result partial segments which themselves contain points are troublesome.

[54]Ibid., p. 83.

rather, fatigues and discourages the mind which cannot see the end of its journey, hidden in a uniformly cloudy region.[55]

The Empiricist thus claims that mathematics has no need of exactitude of measurement at an infinitesimal level, and further, infinite division serves no purpose other than to fatigue and discourage us. Each measurement can be as exact as we need it to be, the division can be carried on until we reach that useful exactitude, and while the mind can perhaps imagine, in some sense, infinite division, nothing practical is to be gained by positing the result of such infinite division. Thus, there are only finite divisions to be made, and only finitely many points on a geometrical line. "[For the Empiricist,] the set of points as dense as one would wish always remains a finite set."[56] The Empiricist thus rejects the original premise of the Idealist's argument – the infinite division of the unit length – and rejects the existence of infinitesimals themselves.

5.5. Continuity and a Unified Mathematics

As the task of this chapter is to find out not only the foundations of mathematics for Paul du Bois-Reymond, but specifically du Bois-Reymond's theory of continuity, we must end this chapter by turning our attention to answering this question. This last section will contain two subsections. In the first, I will present the conception of continuity entailed by the Idealist and Empiricist systems (briefly, the Empiricist has no conception of continuity, while the Idealist does, but that his conception must differ markedly from the continuity of Dedekind and Cantor). In the second, I will discuss du Bois-Reymond's neutral system of mathematics, a system acceptable to both the Idealist and Empiricist tendencies within ourselves; and then I shall briefly discuss the status of the mathematical continuum in this neutral system.

5.5.1. Idealist and Empiricist Continua. Though in the *General Theory of Functions* and in this chapter the Idealist has typically been given the podium first, it is useful here to begin with the Empiricist conception of mathematical continuity, due to its simplicity. In a word, it does not exist. There is

---
[55]Ibid., p. 92.
[56]Ibid., p. 162.

no such thing as mathematical continuity. "For the Empiricist the continuum of numbers does not exist, either as a limit, nor as the final end of an approximation. It absolutely fails to exist for him."[57] Given the finite nature of the set of points on any line segment discussed above, it should not be surprising to realize that the Empiricist also does not believe that the line itself is continuous. The set of numbers, too, contains as many as we wish, but not actually infinitely many; thus, it is unsurprising that a set of numbers cannot reach continuity. This is quite in line with the Empiricist worldview; continuity, like infinity, is one of those idealized concepts of which we never have direct experience, nor can we deduce it from or reduce it to our experience.[58]

The Idealist does have a concept of mathematical continuity, as well as a conception of a continuous straight line. This conception of mathematical continuity is more or less the one we are used to seeing in mathematics, that is, a continuity which is infinite, dense, and has no gaps, though it does have some unusual features. First of all, the Idealist's continuity can never be reached through a foundational approach; that is, the Idealist admits that we cannot begin with empirical facts, such as stones which are counted and organized, and eventually derive a mathematical continuum. For the Idealist, "the continuum of numbers is not the actual limit of any empiricist series of representation."[59] The continuum is divorced from empirical representation, even for the Idealist, who is used to making leaps from representation to idealization.

The second unusual feature of the Idealist's numerical continuum is that does not have the property of Dedekind continuity. As we proved in Chapter 3, Dedekind continuity is inconsistent with the existence of infinitesimals, and, as we saw above, the Idealist believes infinitesimals follow with necessity from the most basic assumptions about the continuity of a straight line. Thus, while it looks as though the Idealist is forwarding a type of Dedekind continuity

---

[57]Ibid., p. 163.

[58]Interestingly, as mentioned elsewhere, Immanuel Kant believed that space and time were continuous, and, given that space and time, according to Kant, were conditions of our experience of reality, under transcendental assumptions we cannot help *but* experience continuity. It seems as though under these assumptions, the Empiricist is quite wrong in his contention that we never experience continuity; rather, we always experience it. Peirce, too, as we shall see next chapter, believed that continuity underlies all human experience and a thorough understanding of our world must always include an understanding of continuity itself. For Peirce as well, continuity seems to be a part of human experience.

[59]Ibid., p. 163.

at the beginning of his proof of infinitesimals, the fact that he goes on to use that conception to argue for the existence of infinitesimals cues us in to the non-Archimedean, and therefore non-Dedekind nature of the continuum. Recall that du Bois-Reymond's Idealist began his argument for the existence of infinitesimals with the sentence, "The proposition that the number of points of division of the unit length is infinitely large produces with logical necessity the belief in the *infinitely small*."[60] The nature of continuity, according to Dedekind, is that wherever I may divide a set, I always do so at a member of the set, never in between members. Wherever a line is divided, it is divided at a point; wherever the set of real numbers is divided, it is divided at a real number.

As this is the very property of Dedekind's real numbers which led to our proof of the Archimedean principle, it must be that the Idealist's infinite division points of the unit length do not directly entail this property. Thus, while there are infinitely many points of division – points marking actual divisions of the unit segment – we must conclude that there is at least the possibility of divisions occurring which do not occur at particular points. Though it seems at first glance that this type of continuity is no continuity at all – that is, that these theoretical divisions which do not occur at division points must therefore occur at gaps in the line or in the number line – this is not necessarily so. For a gap in the line to be proven under such a scenario, the proof would depend upon the assumption that there was nothing on the line but points, and thus, a place which lacked a point would not be part of the line. However, the Idealist clearly believes that there is more to a line than simply points; in fact, he believes that the line decomposes into points *and intervals*.

These two properties of the Idealist's continuum – that continuity can never be reached as a limit of an empirical series, and that continuity is non-Archimedean and lacks Dedekind continuity – are consistent with the Empiricist's indication that continuity is an idealized property which goes beyond direct experience. This is one conclusion both the Empiricist and the Idealist agree upon; they only differ in their willingness to accept mathematical entities so divorced from empirical reality.[61]

---

[60]Ibid., p. 73.

[61]Do note, however, that neither the Idealist nor the Empiricist has claimed that reality itself is continuous, or that it is discontinuous. They only agree in the limits of our ability

It is worth repeating the fundamental feature of magnitude on which du Bois-Reymond depends so heavily for his proof of infinitesimals: every part of the continuum must resemble every other part, in every respect other than size. We saw above what du Bois-Reymond's Idealist made of this feature; we shall see in the next chapter how Peirce uses this same feature of the continuum to great effect.

5.5.2. A Neutral Mathematics. Du Bois-Reymond believed that these two world-views about foundational matters were so diametrically opposed that they were separated as if by a large, chasm, and this chasm "is too profound and too vast to be able to be filled by reciprocal concessions. The counter-proposals are absolutely irreconcilable."[62] Yet, both systems are "equally authorized to serve as the base of the science [of mathematics]."[63] Thus, compromise is impossible, and deciding once and for all in favor of one worldview over the other is also impossible. However, du Bois-Reymond did attempt to create a system that was *objectionable* to neither the Idealist nor the Empiricist. The first step to creating this system is to follow the Empiricist in ridding mathematics of any metaphysical entity underivable from experience. Thus, any infinitely large or infinitely small magnitude will be treated as if it were "an idealist fiction."[64] Similarly, phrases such as "precise measurement" will be replaced with phrases such as "as exact as we wish." The logic of this neutral system, however, will follow that of the Idealist, for restricting ourselves to prove that certain mathematical properties only hold for sets "as large as we wish" will not satisfy the Idealist, and any proof that holds for infinite sets, whether or not these infinities actually exist, will also hold for the weaker sets which are only as large as we wish them to be. Thus, concludes du Bois Reymond, "empiricist language, idealist proofs."[65]

It is surprising how much of mathematics can be done with a system such as this, which is restricted on both ends; ontologically restricted to only objects related to direct experience, proofs restricted to only those strong enough to

---

to experience continuity empirically, just as they agree that whether reality is infinite or not, we can never experience actual infinity.

[62]Ibid., p. 128.
[63]Ibid., p. 129.
[64]Ibid., p. 129.
[65]Ibid., p. 130.

capture actual infinity and perfectly exact measurement. Walking the Idealist/Empiricist tightrope is possible when we are dealing with calculations and procedures that do not address troublesome issues. For example, we can perform the operations of differential calculus without ever settling the question of whether differentials are actual infinitesimals, as Leibniz was sometimes prone to believe, or whether they are simply as small as we would wish. However, vast areas of mathematics are closed to us on this neutral system. Infinitesimals and transfinites are denied to us, but so are actual infinities (and thus, one would imagine, the Axiom of Choice among some other set theoretical results), as well as actual continua.

Notably, du Bois-Reymond's Infinitärcalcül is similarly off bounds in both the Empiricist and the neutral system. Recall from Section 5.2 above that the Infinitärcalcül was a system open to us once we had access to the levels of infinity derived from Idealist principles, varying magnitudes of the infinitely large as well as varying magnitudes of the infinitely small. I do not believe that du Bois-Reymond wished to claim his own mathematical system was out of bounds; but I also do not believe that du Bois-Reymond wholeheartedly accepted this neutral system. It is surprising how much mathematics can be accomplished without the Empiricist/Idealist divide breaking into open warfare, and the neutral system of mathematics ensures that as far as we are able, we can accomplish many standard mathematical calculations without settling the larger foundational issues.

These larger issues, however, must sometimes be confronted, and unfortunately, according to du Bois-Reymond, they are not issues which can be settled; the Idealist/Empiricist divide remains. "Since no conclusive mathematical consideration will ever decide between the two, such questions constitute undecidable problems whose solutions will remain forever outside the range of our mathematical abilities."[66] Among these undecidable issues are the very problems this book is concerned with: whether the set of real numbers is continuous and whether infinitesimal quantities exist.

---

[66] See McCarty [2004, p. 526].

CHAPTER 6

Charles Sanders Peirce

6.1. Introduction

Charles Sanders Peirce was born in Cambridge, Massachusetts, in 1839. He studied science at Harvard, and after receiving his Bachelor of Science degree in chemistry at 1863, went to work for the U. S. Coast and Geodesic Survey program run by his father. Though the extent of his academic appointments was limited to five years as an instructor at Johns Hopkins University, throughout his life he applied himself to studying and writing on a variety of subjects, notably philosophy, logic, mathematics, psychology, and semiotics (the study of signs and signage). Peirce was a prolific writer. Plagued by trigeminal neuralgia, a progressive nerve condition causing severe pain, he was driven to write with great fervor in an attempt to record all of his thoughts before he was no longer capable of intellectual work. Peirce also suffered from financial instability for much of his life; for several years he was financially dependent on the charity of friends, especially William James (1842–1910).[1]

Peirce is known for the development of the American school of Pragmatism, and his philosophical work was heavily influenced by Aristotle, Immanuel Kant, and Georg Cantor. He took a systematic, global approach to philosophy, but also paid much attention to the mathematical and logical implications of his philosophical ideas, developing his own system of logic in order to gain a higher level of precision. Continuity was, for Peirce, an idea which linked his systematic approach to philosophy with his fondness of mathematics. He believed it was essentially a mathematical concept, or at the very least could best be defined

---

[1]For more details of Peirce's life, see Robert Burch's article, "Charles Sanders Peirce," [Burch, 2006] and Joseph Brent's book, *Charles Sanders Peirce: a life*, [Brent, 1993].

and understood using mathematics; and yet he believed continuity could be found in every realm of life, from psychology to history, from philosophy to biology.

Peirce's philosophy of continuity is worth considering in detail for two main reasons. First, like Paul du Bois-Reymond, he believed that infinitesimals were essential to our understanding of continuity, though the way in which he conceived of them differed markedly. Second, he began his investigation into the continuous from a stance quite similar to that of Cantor and Dedekind, though he did include infinitesimals in his conception from the beginning. He ultimately became frustrated with his Cantoresque definition, and ultimately rejected it in favor of a definition which more closely satisfied his intuitions. Though his theories of continuity developed gradually and shifted subtly over many years, we can group them into three general definitions, which I shall call "early," "middle," and "late." In this chapter, I will first briefly introduce Peirce's theory of synechism – the theory that continuity is central to many intellectual disciplines – and then explain each of Peirce's three theories of continuity in turn, including his reasons for rejecting the first two theories. Lastly, I will discuss the interesting relationship between Peirce's infinitesimals and his different theories of continuity.

6.2. Synechism

Though Peirce changed his mind about many things during his long philosophical career, a consistent feature of his philosophy was his theory of synechism. Philip Wiener defines synechism concisely in a footnote. "Synechism is the principle that continuity prevails in all thought, in the evolution of life and of human societies and institutions as well as in the logic of science."[2] Peirce held that while the universe was, in essence, logical, this logical development contended with three separate ontological categories; these categorical absolutes were chance, evolution, and continuity. The universe at every moment has a chance element in it (this is the doctrine of tychism, the theory that absolute chance is always a factor), and the universe tends toward growth (which Peirce

---

[2]See Wiener [1966, p. 187, editor's note].

6.2. SYNECHISM

called agapism, a type of evolutionary love).[3] The nature of the universe can be comprehended when synechism, the omnipresence of continuity, is added to the mix: "if all things are continuous, the universe must be undergoing a continuous growth from non-existence to existence."[4] These three ontological categories together give us a picture of a universe which is constantly coming into being.

These categories affect everything in the universe, not simply the universe as a whole; all natural systems are thus prone to chance factors, tend toward growth, and move in continuous fashions. This helps explain some of the structural similarity we see between systems as diverse as a human life and the Roman Empire; both are subject to these basic ontological categories.[5] However, while tychism and agapism seem to be phenomena in the physical as well as metaphysical world – that is, actual complex systems, such as the Roman empire and the evolution of mankind, exhibit chance elements and a tendency toward change in actuality – synechism also affects the most abstruse of natural systems, the human mind. In "The Law of Mind," he introduces synechism by writing that "the next step in the study of cosmology must be to examine the general law of mental action."[6] Ideas, Peirce writes, "tend to spread continuously and to affect certain others which stand to them in a peculiar relation of affectability. In this spreading they lose intensity, and especially the power of affecting others, but gain generality and become welded with other ideas."[7]

---

[3] These three ontological categories are summarized admirably in Feibleman [1946, p. 404–410]. Peirce's thought on all three is discussed carefully in a series of articles written for the *Monist* in 1891–1893, including "The Law of Mind," addressing synechism, and "Evolutionary Love," addressing agapism. The five articles are reprinted in [Peirce, 1998b] and also in [Peirce, 1992].

[4] CP 1.175. References to [Peirce, 1960], *Collected Papers of Charles Sanders Peirce*, (CP) will follow standard notation rather than denote page number.

[5] Peirce has a fascinating discussion in "Evolutionary Love" [Peirce, 1893] in which he postulates that 33 is an important number of years for a human life, as 33 is the rough age at which most men reproduce, and thus, so too, historical movements ought to have a rough time period after which they are naturally (though not always) supplanted by another historical movement. He traces the key dates of the Roman Empire, finding a significant change in direction after every 500 years or so; he finds similar intervals of time taking place between various eras of human thought. See Peirce [1998b, p. 295–296].

[6] See Peirce [1992, p. 313].

[7] Ibid., p. 313.

Individual ideas only have unique identities in a momentary, mental-state sense; an individual consciousness experiences an event, an "idea," and when the moment is over, the idea is gone, mutated, developed, or simply absent. Any attempt to think of the same idea again is a different individual idea, a different state of consciousness, though the content may be at least roughly the same. The connection between these ideas – the past, original formulation of an idea, and my recalling or rethinking a similar idea later – must consist of a relationship between individual ideas. Furthermore, argues Peirce, cognition necessarily takes place in time, and distinct ideas are separated by actual infinitesimal interludes. Thus, continuity affects every aspect of the universe, but most particularly, human thought itself.

Synechism demanded that Peirce study multiple academic disciplines, but he relied upon the tools of mathematics and logic to define continuity itself. In short, Peirce agreed with Aristotle about the prevalence of continuity in the physical world, and he agreed with Dedekind about the ability of mathematics alone to provide us with the tools necessary to build a theory of continuity. His early and middle definitions of continuity are quite mathematical in nature. Even his final definition of continuity, which breaks away substantially from continuity as a primarily mathematical or logical idea, depends upon mathematical procedures and examples for its explication.

6.3. Early Definition of Continuity

In 1878, Peirce defined continuity as "the passage from one form to another by insensible degrees."[8] This definition was presented in the course of Peirce's theory of the continuity of botany – how two very similar examples of the same botanical species differ from each other by insensible degrees, by the slightly different shape of a particular leaf or slightly different coloration in a marking. Dissatisfied with the imprecision of this definition, in 1893 he amended it with a footnote, writing that this account of continuity was meant to refer to "limitless intermediation, i.e., of a series between every two members of which there is another member of it."[9] Though by 1893 Peirce had already developed his middle theory of continuity, he understood his earlier theory to be equating

---

[8]CP 2.646.
[9]CP 2.646, footnote added in 1893.

continuity with density: between any two members of a series, another member can be found. However, the set of rational numbers has the property of density, and the rationals are clearly non-continuous;[10] thus, he realized, this first definition was inadequate. In fact, in his 1893 footnotes to this passage, Peirce suggested that to save the integrity of this early essay, the word "continuity" be replaced throughout in this essay with "limitless intermediation." It is clear that by the time he added these explanatory footnotes, he had rejected density as a sufficient condition for continuity.

6.4. Middle Definition of Continuity

Peirce retained density as a necessary condition for continuity, however. In 1893, in "The Logic of Quantity," he laid out what he then saw as the three conditions for continuity: non-denumerability, Kanticity, and Aristotlicity.[11] The first condition, non-denumerability, requires that if a set is continuous, it (i) must be infinite, and (ii) cannot be in one-to-one correspondence with the natural numbers. The second condition was dubbed 'Kanticity,' to give credit to Kant as the inspiration, and is the property that "between any two points upon [the line] there are points."[12] This, of course, is just the density requirement of Peirce's early definition of continuity already rejected as inadequate by itself. He notes its insufficiency, and thus the need for the other two requirements, by showing that Kanticity holds true of a line from which a closed segment has been removed – a line with a clear gap in it, which is certainly not continuous.[13] Thus, "the union of the disjoint line segments $(-\infty, a)$ and $[b, \infty)$, $a < b$, is dense and, thus, continuous according to Kanticity."[14] Note that such a line, with an obvious gap and thus noncontinuous, also satisfies condition one,

---

[10]Recall from previous chapters, the set of all rational numbers is typically not viewed as continuous as there are demonstrable gaps in the set, e.g., wherever an irrational number would fall. As we shall see below, Peirce eventually believed that the set of all real numbers could not be continuous, either, as there were greater magnitudes of infinity, and thus, the real numbers too were missing something, which a continuum must not.

[11]In section 6 of "The Logic of Quantity," entitled "The Continuum." CP 4.121–124. Peirce forwarded a similar definition of continuity in 1892 (CP 6.121–124) and again in 1903 (CP 6.166).

[12]CP 4.121.

[13]CP 4.121.

[14]See Gwartney-Gibbs [2007, Section 1].

nondenumerability. Clearly another condition is needed to prevent such non-continuous sets from being captured by our definition of continuity.

Peirce's called his third condition of continuity "Aristotlicity" as an explicit nod to Aristotle as the inspiration. A set has Aristotlicity when it is the type of set "whose parts have a common limit."[15] Peirce would seem here to be referencing the tripartite distinction between the various modes in which two things can be next to each other, which was discussed in Chapter 1, above. Aristotle's tripartite distinction appears in the *Metaphysics*, 1069a5 (which Peirce cites in a footnote), and again in the *Physics*, Book V chapter 3, where Aristotle distinguished among various ways in which two things can be considered together: they can be *in succession*, they can be *contiguous*, or they can be *continuous*. The distinction in the *Physics* occurs as part of his argument that continua are not composed of indivisibles.[16] To recount briefly: two things of the same kind are in succession when no third thing of the same kind is between them (but something of a different kind may be). Two things that are contiguous are in succession, and also, their external borders touch, so that nothing is between them, either of the same kind or of a different kind. As for continuity,

> The continuous is a subdivision of the contiguous: things are called continuous when the touching limits of each become one and the same and are, as the word implies, contained in each other: continuity is impossible if these extremities are two. This definition makes it plain that continuity belongs to the things that naturally in virtue of their mutual contact form a unity.[17]

This version of Aristotle's continuity requires, in addition to infinite divisibility, that all limits or borders are shared in common.

By Aristotlicity, however, Peirce did not literally mean subsets of continuous sets must "touch" or "share borders." Vincent Potter and Paul Shields defined

---

[15] CP 4.121.
[16] Aristotle, *Physics*, 227a10–15.
[17] Ibid. 227a10–15.

Peirce's Aristotlicity as "the requirement that a continuum contain its limit-points."[18] Peirce himself later, in 1903, defined Aristotlicity as "having every point that is a limit to an infinite series of points ... belong to the system."[19] And similarly, in 1892, he stated that Aristotlicity holds "if a series of points up to a limit is included in a continuum the limit is included."[20] Thus, a set A has the property of Aristotlicity if and only if the limit of every subset of A is included in the set A itself. This seems related to Aristotle's definition of continuity in terms of parts sharing borders; the borders here would be analogous to the limits of the subsets. However, the analogy can only be a loose one; if we were to take it literally, the requirement of Aristotlicity would lead us to believe that every segment on a continuous line must share end points with every other segment, which is clearly false.[21]

Indeed, Peirce's requirement of Aristotlicity seems more closely related to Cantor's notion of perfect sets than it does to Aristotle's definition of continuity.[22] In fact, Peirce was explicit about the extent to which Cantor influenced his definition,[23] and Potter and Shields compared Peirce's Aristotlicity to Cantor's requirement that a continuous set must be perfect.[24] Recall from Chapter 4 that for a set P to be perfect, it must be equal to each of its derived sets $P^{(\gamma)}$. Thus, a perfect system is equal to its first derivative, and "the first derivative of a system is simply its collection of limit-points."[25] Cantor's perfection implies that every limit-point of a system is contained within the system – which is Aristotlicity – but it also implies that "every point of the system is a limit-point of the system."[26] Aristotlicity is thus half of Cantor's perfection.

---

[18] See Potter and Shields [1977, p. 25].

[19] CP 6.166; written in 1903 in the margin to his 1889 article in his personal copy of the *Century Dictionary*.

[20] CP 6.122; written in 1892.

[21] However, do note that two segments on a continuous line which are next to each other do indeed share an end point.

[22] The degree to which Peirce was directly influenced by Cantor on this and other questions is one of some historical interest. Matthew Moore thoroughly investigates Peirce's interactions with and access to Cantor's seminal works and draws several conclusions of pirority in Moore [2010].

[23] CP 4.121.

[24] Potter and Shields, p. 23.

[25] Ibid., p. 24.

[26] Ibid., p. 24.

With all three conditions in place, we have that a set A is continuous if and only if:

i) it is infinite and non-denumerable,
ii) it is dense, and
iii) for every subset of A which has a limit, that limit is itself a member of A.

Sets with clear gaps, such as the union of the disjoint line segments $(-\infty, a)$ and $[b, \infty)$, $a < b$, considered above, are not continuous according to this definition: the third condition would require that a itself, as a limit of the subset $(-\infty, a)$, be included in the set. If a were included, then the set would instead be the union of $(-\infty, a]$ and $[b, \infty)$, but since $a < b$, the set would no longer be dense, as a and b are both in this set yet there is nothing between them. The set of rational numbers is also excluded from continuity by the third condition; as we saw Chapter 3, Dedekind proved that some sets of rational numbers have non-rational limits. In fact, Peirce's middle definition is rather like Dedekind's definition of continuity in several respects, and as we shall prove in the section on infinitesimals below, Peirce's definition is logically implied by Dedekind's.

This middle definition is adequate in the sense that it excludes clearly non-continuous sets such as the set of rational numbers and discontinuous line segments. Peirce was satisfied with it for several years, but he ultimately rejected it in favor of a third and final definition.[27] The reasons for this rejection are complex and highly philosophical in nature, but they stem from two sources: first, Peirce's re-interpretation of Kant's definition of continuity, and second, a concern related to the infinitely many magnitudes of infinity which Peirce came to recognize after he had developed this second, Cantorian definition of continuity.

---

[27]It is, as we have noted above, an over-simplification to say that Peirce only had three definitions of continuity. Indeed, his supermultitudinous continuity, considered in Section 6.5 below, changed and mutated over the years, as did the previous two definitions. For the sake of consistent exposition, we will here restrict ourselves to considering a thoroughly developed exemplar of Peirce's supermultitudinous continuity, rather than trace the ways in which this mutated and shifted over time.

6.5. Later Definition of Continuity

By the late nineteenth century,[28] Peirce rejected this definition for two main reasons. First, he had come to believe that he had misunderstood Kant's definition of continuity, and sought to rectify this mistake. Second, as he realized the real numbers did not represent the highest level of infinity, he came to believe that they could not possibly be continuous, and thus, any definition of continuity which defined the reals as such was inadequate. I shall address both of these changes of heart in turn.

The density requirement of Peirce's middle theory had been named Kanticity because Peirce believed that Kant's definition of continuity at A169 (B212) of the *Critique of Pure Reason*[29] was equivalent with density. A169 reads, "The property of magnitudes by which no part of them is the smallest possible, that is, by which no part is simple, is called their continuity."[30] It is possible to understand this property of magnitudes as a form of infinite divisibility, which, with respect to numbers and points, could be rendered as density, but in his later work, Peirce came to reinterpret this as meaning that a continuum was something such that "All of [its] parts have parts of the same kind,"[31] that is, every part of the continuum resembles every other part, except with respect to size itself. For Peirce, this no longer implied that between any two points there is a third point; rather, "Kants real definition implies that a continuous line contains no points."[32] If one believes that a continuum must be infinitely divisible in the sense that no part of it is indivisible, then an indivisible point cannot be an essential part of the continuum. In fact, wherever a point appears, according to this third definition of continuity, that point "interrupts the continuity."[33] To remain true to continuity itself, any division of the continuum must yield parts that resemble precisely every other part, no matter how big

---

[28]Peirce forwarded something very much like his third definition of continuity as early as 1895, but in 1903, he repeated his middle definition of continuity. By 1905, however, he had completely forsaken his middle definition in favor of his later one.

[29]See Peirce [1892, p. 320 f. 10].

[30]Kant, A 169 (B211).

[31]CP 6.168. This was written in 1903.

[32]CP 6.168.

[33]CP 6.168.

or small, and while points are useful for navigating on the line, and measuring with precision, they cannot be essential parts of it.[34]

Thus, Peirce abandoned his middle theory of the continuum in part due to his new-found understanding of Kant's definition. In a not unrelated change of heart, he posited that a continuum must have room for as many members as possible; as he became aware of infinitely many magnitudes of infinity, he came to believe that a true continuum could not be limited to any particular magnitude. Thus, not only was it impossible for the real numbers to form a continuum, but definitions of continuity within standard set theory would likewise be impossible, as all sets, no matter how large, have a particular magnitude. Peirce was considering the theory of non-denumerability and its implications as early as 1893, though he had not yet satisfied himself as to how many magnitudes of infinity there were. He defined three classes of magnitude: those that are enumerable, i.e., finite; those that are denumerable, i.e., countable and infinite; and those that are non-denumerable, i.e., infinite and uncountable. He then wrote:

> [These] three classes of multitude seem to form a closed system. Still, nothing in those definitions prevents there being many grades of multiplicity in the third class. I leave the question open, while inclining to the belief that there are such grades.[35]

He later proved there were infinitely many such grades, and hence rejected the thesis that the set of real numbers was continuous, as there must be infinitely many magnitudes greater than this set.[36] By this point in his life, Peirce firmly believed that a continuum must allow room for all possible points, therefore, there could not possibly be a collection larger than one which is continuous.

Peirce's insistence that true continuity must be in a sense absolute plenum, there must at least be room for all possible sub-entities within a continuous object, seems to be at least partly derived from an examination of our experiences of continuous phenomena, such as time. As he wrote in an 1895 essay,

---

[34]Notice that this has similarities to du Bois-Reymond's conception of continuity, as described in our previous chapter.

[35]CP 4.121.

[36]See [Peirce, 1976, v. 3, p. 880, footnote 2].

"On Quantity, With Special Reference to Collectional and Mathematical Infinity,"[37] humans do seem to have sense experiences of instants, such as the present instant, but humans also perceive the flow of time. This perception of flow cannot, Peirce states, accurately be captured by "a series of instantaneous photographs ... no matter how closely they follow one another,"[38] and so too, it cannot be captured by a series of instants, no matter how close. The continuity of time must either be characterized by any length of time having "room for any multitude, however great, of distinct instants," or by viewing the instants as "so close together as to merge into one another, so that they are not distinct from one another."[39] Neither conception of continuity can be satisfied if the number of instants is limited to the cardinality of the real numbers; this cardinality does not necessitating a merging of instants, nor would this cardinality be able to contain "any multitude, however great," as there are greater multitudes than this.

The same holds true of the continuous straight line. As Peirce wrote in 1908, he perceived the line as continuous in the same sense as a segment of time, stated above:

> [T]here is room on a line for a collection of points of any multitude whatsoever, and not merely for a multitude equal to that of the different irrational values, which is, excepting one, the smallest of all infinite multitudes, while there is a denumeral multitude of distinctly greater multitudes, as is now, on all hands, admitted.[40]

The continuous line must have room for any cardinality of points whatsoever, not simply the cardinality of the real numbers; and thus, the Cantor-Dedekind axiom, that the points on a line and the real numbers are in correspondence, must fail. Further, any particular cardinality of points, no matter how large the transfinite number we use to represent it, must also fail to serve as the location for continuity, as no particular transfinite is the largest one; we can always generate one larger. In order to support this possibility for a continuum to

---

[37] See Peirce [1976, v. 3, p. 39–63, especially §5, p. 58–63].
[38] Ibid., p. 59.
[39] Ibid., p. 60.
[40] Ibid., p. 880, footnote 2.

have room for "any multitude, no matter how great" of instants or points, one must move away from generating the continuum from any particular multitude. In the quotation above, Peirce is clearly assuming the continuum hypothesis - that the collection of real numbers is equivalent to the smallest nondenumerable magnitude of infinity. If one assumes the continuum hypothesis, and also that true continuity must leave room for a collection of individual entities of any magnitude, the Cantor-Dedekind axiom must fail and the line, if it is continuous, must be infinitely richer than the real numbers. The continuum hypothesis in particular, however, is not a necessary assumption in this argument; all that is required is that we view the collection of real numbers as some particular magnitude of infinity or another, as Cantor and Peirce have both proved that given any particular magnitude of infinity, we can use it to construct the next greatest magnitude simply by taking the power set of the magnitude in question. Thus, wherever the collection of real numbers falls on the line-up of transfinite magnitudes, it can never be the largest one, and thus, a continuous line in Peirce's sense will be infinitely richer than any particular set of numbers, by its capacity of containing points of any magnitudes whatsoever.

It follows from this that continuity cannot be composed of numbers or points. For if there were a set made up of numbers or points, it would be a multitude with a specific magnitude, and there would be at least one magnitude greater than it. Therefore, "we must either hold that there are not as many points upon a line as there might be, or else we must say that points are in some sense fictions which are freely made up when and where they are wanted."[41] Peirce chose the latter option. He also argued that continuous lines could not be made up of points because there will always be gaps between the points. Thus, "there is no reason why the points of one line might not slip through between those of the other. The very word *continuity* implies that the instants of time or the points of a line are everywhere welded together."[42] Dedekind in particular, Peirce wrote, presented a continuum that was not actually continuous, and was deeply flawed by its insistence that continua could be composed of individual points or numbers. At this point in his life Peirce

---

[41]Ibid., p. 58–59.
[42]Ibid., p. 60–61.

believed that "Kant's ideas about the continuum, and Dedekind's further elaboration thereof, are faulty ... because they use a foundationalist conception of a line as consisting of points."[43]

This, then, was Peirce's solution and final definition of continuity – an intuitive, "true" continuum must consist of points so close to each other that they are literally welded together.[44] There would have to be so many elements in this continuum that it is larger than any given multitude – and thus would be too large itself to be a multitude. Peirce called this a *supermultitudinous* collection, but do not be fooled by the inclusion of the word "collection," for this continuum is no longer a collection of discrete elements:

> A supermultitudinous collection, then, is no longer *discrete*; but it is *continuous*. As such the term "multitude" ceases to be applicable to it.[45]

Supermultitudinous collections, Peirce wrote, are created by starting with a non-denumerable multitude, such as the real numbers. Then, between any two numbers, all of the real numbers between zero and one (the collection of which is itself a non-denumerable multitude) are inserted. If one continues to insert non-denumerable multitudes between all the numbers, Peirce claimed, the numbers themselves will eventually lose their distinction and become a supermultitudinous collection:

> I am careful not to call supermultitudinous collections multitudes. Multitudes imply an independence of the individuals of one another which is not found in supermultitudinous sets. Here the elements are cemented together, they become indistinct.[46]

---

[43]See Herron [1997, p. 604].

[44]Peirce was not the first person to conceive of continuity in this manner. Galileo (1564–1642), who was himself troubled by the inability of himself and his peers to mathematically define continuity, believed that continuous magnitudes were composed of indivisibles, but that "the aggregation of these [indivisibles] is not one resembling a very fine powder but rather a sort of merging of parts into unity, as in the case of fluids" [Boyer, 1959, p. 116].

[45]See Peirce [1976, v. 3, p. 87].

[46]Ibid., v. 3, p, 87–89.

Numbers would "cement together," as would points. Any distinct, individual point would disrupt this continuity. As Hausman wrote: "A point marks a break in a continuum, and a collection of points is as much a collection of gaps as a collection of units."[47]

While one might potentially imagine instants of time melding into one another, depending on one's conception of the metaphysics of an instant, it is difficult to imagine numbers melding into one another. If we actually inserted the set of real numbers between one and zero between any two numbers on the line, the result would be equinumerous to the set of real numbers that we began with, as that is what it means to say the real numbers are dense – there are infinitely, and in fact uncountably, many numbers between any two numbers. Adding uncountably more would not increase the magnitude of the set of real numbers, nor would it force the numbers to collapse upon each other. There would simply still be uncountably many.

Also disturbing is Peirce's tendency to wish to compose this "true" continuum from points (or numbers) added to each other until they weld together, and simultaneously to claim the line has no points (or numbers) at all. Furthermore, while accusing Dedekind of having a compositional view of continuity, he composed his own continuity from points. The best we can do for Peirce at this point is to assume that this "true" continuum is not formed from actually melding points, but that rather, the compositional story he has told must be more of tool to help us think of what a line without points would be like, in spite of Peirce's own insistence that this procedure is a creation of some sort.

Even if we save Peirce from the hypocrisy of blaming Dedekind for having a compositional continuum while at the same time composing his own continuum out of points, this third definition of continuity has further problems, problems which Peirce himself could see. We shall discuss some of these difficulties in the next section.

---

[47]See Hausman [1998, p. 630].

6.6. Advantages and Disadvantages of Peirce's Late Continuity

Peirce believed that the greatest advantage of this third definition of continuity was that it provided a continuum that lived up to his intuitive notions of continuity, i.e., absolute smoothness. When one divides such continuity, every part resembles the whole, and the whole resembles every part. Peirce disliked Dedekind's cuts because they violate this latter intuitive idea about continuity. As Herron wrote, "Peirce is fond of producing gaps in the continuum as one way to intuitively convince the reader that Dedekind's definition of a continuous line is not sufficient to define a true continuum."[48] Peirce also thought the asymmetry of Dedekind's real numbers told against their continuity. When one cuts the rationals at a place where no rational number creates the cut, one creates an irrational number, and then adds it to one side or the other to complete the cut. This means that every Dedekind cut is an asymmetry – one side has an end point while the other does not.[49]

Peirce was determined to find a definition of continuity that satisfied his intuitions, because intuitive continuity is important to the theory of synechism. Recall that synechism is the theory that continuity can be found everywhere, in every physical object, in every intellectual theory. The sort of continuity Peirce saw in the history of the Roman Empire was not a continuity that could easily fit into a mathematical mold. He believed firmly that continuity was smoothness that went beyond conceptions of number and point, and he eventually created, with this final definition of continuity, a theory that might possibly have lived up to his intuitions.

The cost of retaining his intuitions was great. The main disadvantage of Peirce's intuitive continuum is that one cannot use it in mathematics. In essence, by rejecting a Cantorian continuum in favor of a philosophical model that satisfied his intuitions, Peirce rejected mathematical continuity altogether. This new continuum, existing only in potentiality, with no numbers and only potential points, is difficult to use in calculations. Supermultitudinous collections themselves, assuming they could be consistently defined, go beyond what

---

[48]See Herron [1997, p. 606–607].
[49]See Peirce [1976, v. 3, p. 93].

Peirce thought was useful for mathematics. In speaking of the *second abnumeral multitude*, which, if one assumes the Continuum Hypothesis to be true (as Peirce does), would be the power set of the real numbers, Peirce claimed:

> Mathematics affords no example of such multitude. Mathematics has no occasion to consider multitudes as great as this.[50]

This meant, for Peirce, that continua such as space, time, and the straight line, were larger and more complex than any system of numbers could reach. No matter how large, no set of numbers could be used to measure space, time, lines, geometrical figures, or, one supposes, objects that exist in space or time:

> When the scale of numbers, rational and irrational, is applied to a line, the numbers are insufficient for exactitude.[51]

Peirce then described points as the "hazy outlined part of the line whereon is placed a single number," not precise determinations of exact measurement.[52] However, if measurement fails, so too does every part of mathematics where numbers and continua supposedly mix, such as calculus, topology, and all of geometry.

Of course, Peirce's ultimate goal was not to create a mathematical continuity. His goal was to explain the mysterious force that he saw linking all branches of science and physical experience of the world. As Dauben wrote:

> Peirce's interests had never been inclined towards analysis. From the very beginning he had been inspired by the purely logical implications of the syllogism of transposed quantity, and the logic of relations. Thus, unlike Cantor, he was not concerned to develop the arithmetic properties of his ideas ... He was interested in illuminating a deep philosophical problem of long standing, namely that of the continuum, and he felt that

---

[50]Ibid., p. 85.
[51]Ibid., p. 127.
[52]Ibid.

6.7. PEIRCE'S INFINITESIMALS AND HIS MATHEMATICAL CONTINUITY

conceptually he had found an approach to the subject that was the most satisfying of all.[53]

Thus, in some sense, comparing Peirce's final definition of continuity to the mathematical continuum of the other three figures studied in this book is like comparing apples and oranges. Peirce's answer to the question, "How can a mathematical continuum match up to geometry and to the real world" is, quite clearly, "It can't." However, he did provide us with a mathematical continuum in his middle definition of continuity, and this definition, while not representing his fully developed philosophical opinion on the matter, can be compared to those of Cantor, Dedekind, and du Bois-Reymond. In fact, since Peirce himself did not believe mathematics could be done with his "true" continuum, it is possible that he himself would have agreed to the use of his middle definition of continuity for the purposes of doing mathematics.

Thus, in the last part of this chapter, it is essential to consider the theory of infinitesimals that corresponds with Peirce's middle, mathematical definition of continuity, although the theory of infinitesimals which corresponds with his final definition of continuity will also be considered briefly.

6.7. Peirce's Infinitesimals and his Mathematical Continuity

Peirce stated several times that the theory of infinitesimals was preferable to the "cumbrous" theory of limits.[54] In 1892, he gave a definition of infinitesimals designed to demonstrate their superiority to limits. This definition appeared in "The Law of Mind,"[55] along with a discussion of his mathematical continuity. At the time, he still believed that Kant's view "confounds [continuity] with infinite divisibility"[56] and at that time he knew that infinite divisibility

---

[53]See Dauben [1977, p. 131].

[54]For example, he wrote, "Men are afraid of infinitesimals, and resort to the cumbrous method of limits. This timidity is a psychological phenomenon which history explains. But I will not occupy space with that here." CP 4.151. Also see the footnote to CP 4.118: "The doctrine of limits should be understood to rest upon the general principle that every proposition must be interpreted as referring to a possible experience," which was a mathematical position Peirce found inferior to the simple introduction of infinitesimals.

[55]See Peirce [1998a, p. 312–333].

[56]Ibid., p. 320.

(i.e., density) was not sufficient for continuity. Peirce also believed at this time that real numbers and points on a line were more or less interchangeable, and substituted one for the other quite freely. It was during the discussion of the continuity of real numbers in particular that he gave his definition of infinitesimals.

> Every number whose expression in decimals requires but a finite number of places of decimals is commensurable[57]. Therefore, incommensurable numbers suppose an infinitieth place of decimals. The word infinitesimal is simply the Latin form of infinitieth; that is, it is an ordinal formed from *infinitum*, as centesimal from *centum*. Thus, continuity supposes infinitesimal quantities. There is nothing contradictory about the idea of such quantities. In adding and multiplying them the continuity must not be broken up, and consequently they are precisely like any other quantities, except that neither the syllogism of transposed quantity nor the Fermatian inference applies to them.[58]

This is a remarkable claim, one worth dissecting carefully. Starting at the end of the argument, Peirce claimed that neither the "Fermatian inference" nor the "syllogism of transposed quantity" apply to infinitesimals. The Fermatian inference refers to complete mathematical induction, according to Stephen Levy;[59] thus one cannot argue by mathematical induction when one is dealing with a system that includes infinitesimal quantities. The syllogism of transposed quantity was a favorite of Peirce's and he discussed it in many different places, including this 1892 article in the *Monist*, where he explained it in terms of Frenchmen:

> Every young Frenchman boasts of having seduced some Frenchwoman. Now, as a woman can only be seduced once, and there

---

[57]I.e., is rational. The Greeks first discovered irrational quantities by discovering geometric lengths that were incommensurable with a unit length.
[58]Ibid., p. 322.
[59]See Levy [1991, p. 131].

6.7. PEIRCE'S INFINITESIMALS AND HIS MATHEMATICAL CONTINUITY 125

are no more Frenchwomen than Frenchmen, it follows, if these boasts are true, that no French women escape seduction.[60]

For this inference to hold it is not necessary to have two distinct classes (i.e., Frenchmen and Frenchwomen); elsewhere his inference was stated in terms of Texans killing Texans, or Hottentots killing Hottentots.[61] However, no matter how many classes are involved, the inference only holds if the set or sets being referred to are *finite*. If the sets are infinite, the conclusion does not follow, and this fact is precisely what made the syllogism so interesting to Peirce. Thus, if the inference of transposed quantity fails to apply to infinitesimals, it seems that there are infinitely many of them.[62]

Peirce provided us with a concrete example of a number at the infinitieth place, and how that should figure into our mathematical thinking. It is generally agreed upon that the infinite repeating decimal expansion 0.999... is equivalent to the number 1.[63] Peirce disagrees with this equivalence. He argues that they do indeed differ, but only at the infinitieth place. Thus, 0.999... is infinitesimally smaller than 1.[64] One might interpret this as meaning that the expression $1 - 0.999\ldots$ would be equal to 0.000...1, with an infinite number of zeros replacing the ellipses, and thus that 0.000...1 represents a mathematical infinitesimal. One also assumes that 0.333... would differ from 1/3 by an infinitesimal amount, though Peirce did not address this question directly.

Thus his infinitesimal is similar to the mathematical differential. In calculus, a differential is related to linear approximations. Consider a point a on the graph of a known function. Linear approximations (using the derivative of

---

[60]See Peirce [1992, p. 316].

[61]See Peirce [1976, v. 3, p. 772], for the Hottentot formulation, and CP 3.288 for the Texan one. We might wonder why Texans and Hottentots are so violent; however, we might also wonder if the French would agree that each Frenchwoman can only be seduced once.

[62]This is equivalent to Dedekind's definition of an infinite set as any set which can be placed into one-to-one correspondence with a proper subset of itself.

[63]Though it hardly seems necessary, let me briefly prove this. One divided by three is 1/3, which is precisely equivalent to **0.333...** . If we add $1/3 + 1/3 + 1/3$ we get 1; if we add **0.333... + 0.333... + 0.333...** we get **0.999...** Since 1/3 and **0.333...** are equivalent, so too must **0.999...** be equivalent to 1.

[64]See Levy [1991, p. 130]; also see Peirce [1976, v. 3, p. 597]: "although the difference, being infinitesimal, is less than any number can express[,] the difference exists all the same, and sometimes takes a quite easily intelligible form."

the known point and the increment along the x-axis between the known point and the unknown point) can help us approximate the secant line, connecting the known point a with an unknown point x. Of course, the closer a is to x, the closer our approximation can become. If the two points are the same, the calculation falls apart, and growing distance makes it unreliable; the ideal distance is an infinitesimal one.

Peirce thus seems to be claiming that infinitesimals exist as quantities defined by such a differential. Leibniz himself conceived of a differential as a mathematical entity related to the infinitesimal. The main problem with Peirce's infinitesimal is that he introduced it too soon; it was defined by Peirce in 1892, when he still believed his middle definition of continuity was the correct one. However, this definition logically implies the Archimedean principle, which, as we saw in Chapter 3, is mathematically inconsistent with the existence of infinitesimals.

The implication is easy to prove: Peirce's mathematical continuity implies Dedekind's continuity, and we proved in Chapter 3 that Dedekind's continuity implies the Archimedean principle.

THEOREM. *There is a set P which is non-denumerably large, has Aristotlicity and Kanticity.*

PROOF. Assume for *reductio* that P is not Dedekind-continuous; i.e., there is a cut in P for which there is no member of P at which the cut takes place.

Since set P is not Dedekind-continuous, there is a division in P such that every member of subset A is less than every member of subset B, and the union of A and B equals the set P. Furthermore, this set does not occur at a point, which means that there is no greatest member of A nor least member of B. Consider for a moment the subset A. This set clearly approaches the cut, but does not go beyond it; therefore, A has a limit. However, because the cut does not occur at a specific member of P, the limit itself is not a member of P. This violates Aristotlicity, and thus, we have our contradiction. □

6.7. PEIRCE'S INFINITESIMALS AND HIS MATHEMATICAL CONTINUITY

Thus, either Peirce's infinitesimals are inconsistent with continuity as he understood it in this period, or they are consistent with the Archimedean principle. Yet it is simple to show that Peirce's infinitesimals violate the Archimedean principle. In Chapter 3, we used Waismann's definition of the Archimedean principle:

> If a and b are any two positive numbers of [a] system and $a < b$, then it should be possible to add a often enough that the sum $a + a + \cdots + a$ eventually surpasses b. Briefly stated, there should always exist a natural number n such that $na > b$.[65]

Consider our infinitesimal $0.000\ldots 1$ as a and 1 as b. Is it possible to multiply $0.000\ldots 1$ by a number large enough that it surpasses 1? The answer is yes, but only if that number is itself infinitely large. However, the Archimedean principle traditionally requires n to be finite; it requires a finite addition $a + a + \cdots + a$, not an infinite addition. Therefore, Peirce's infinitesimals are inconsistent with his middle definition of continuity. However, this inconsistency only holds if one insists upon viewing infinitesimals as numbers; these infinitesimals can still be a part of mathematics in the same way that differentials can be used in calculations without being themselves members of the set of real numbers.

In 1900, Peirce wrote a letter to the editor of *Science*, defending his position that differentials could be considered to be true infinitesimals.[66] Strangely, rather than defend this claim, he did not discuss infinitesimals at all, but rather ended up introducing his third and final definition of continuity: a continuum which does not contain and is not constituted by points. Perhaps Peirce believed that the introduction of this later definition was in itself a proof of the existence of infinitesimals. By presenting a continuum that has no points at all, he again asserted his belief that infinitesimals are an integral feature of continuity. However, any infinitesimals which are a part of Peirce's later conception of continuity must be very different from the infinitesimals defined above as differentials, and Peirce did not answer the charge that infinitesimals resembling differentials were inconsistent with his middle definition.

---

[65]See Waismann [2003, p. 209].
[66]CP 3.563–570.

It is possible that Peirce himself would answer my charge of inconsistency by explaining that the failure is not in infinitesimals themselves, but in his middle definition of continuity, which he completely abandoned in his later career. After all, in his final theory continuity cannot be Archimedean in nature, as the Archimedean principle requires numbers or points, and his final continuum has neither. Therefore, there seems to be no contradiction between this continuity and the existence of infinitesimals. However, abandoning his middle definition and siding with his later definition would leave us in the position we were in at the beginning of this section; since we cannot calculate with this unwieldy, largely potential continuum, Peirce has ultimately not given us a theory of infinitesimals which can be used in mathematics.

6.8. Peirce's Infinitesimals and his Final Definition of Continuity

All that remains in this chapter is to describe how infinitesimals themselves underwent changes when Peirce's continuity changed, and to outline the role they play in his final theory of continuity, even if that role is mathematically useless. Remember that, according to his third definition of continuity, traditional non-extensional points are impossible, as they interrupt continuity by introducing a part of the continuum that does not mirror every other part. The discontinuity of a point comes from the fact that it does not share the infinite divisibility[67] of the rest of the continuum. However, infinitesimals, for Peirce, are infinitely divisible. As John Bell wrote:

> Now the "coherence" of a continuum entails that each of its (connected) parts is also a continuum, and, accordingly, divisible.[68]

---

[67]Though early in Peirce's career he conflates infinite divisibility and density, at this point, infinite divisibility means just that – the ability to literally divide the continuum or any part of the continuum infinitely many times. As density implies that between any two points there is another point, density clearly does not apply to a continuum without any points.

[68]See Bell [2005].

6.8. PEIRCE'S INFINITESIMALS AND HIS FINAL DEFINITION OF CONTINUITY 129

Of course, when an infinitesimal is divided, the result is another infinitesimal. Peirce clearly thought there were different magnitudes of infinitesimals; how many magnitudes, however, is a mystery.

Thus, numbers and points cannot be used to measure a continuum, as the very definition of this final continuum prohibits any of its parts from being indivisible. However, infinitesimal-points are not forbidden, as they themselves are infinitely divisible, and thus, they can be used to estimate measurements on this type of continuum, although, as we mentioned above, "it [is] intrinsically doubtful precisely where each number is placed."[69] We can, however, identify the near neighborhood of the number as an infinitesimal-point, and thus can approximate exactitude. "Thus a point is the hazily outlined part of the line whereon is placed a single number."[70] This gives us a clue of how we might do something resembling mathematical calculation and measurement using Peirce's unwieldy third continuum: by substituting infinitesimals for points or numbers.

As intriguing as this pseudo-calculation might have been, Peirce did not develop it further. According to Dauben, "Peirce did not undertake a careful arithmetic investigation of the properties of his infinitesimals, nor did he undertake any investigation of non-Archimedean systems in general."[71] And, further, "Ultimately Peirce's infinitesimals remained vague rather than rigorously defined mathematical entities. He never suggested how they might be useful in analysis."[72] Like Paul du Bois-Reymond, Charles Peirce believed in the existence and usefulness of infinitesimals, but unfortunately he did not provide us with a coherent and consistent calculus which included them.

Thus, Peirce's middle and late definitions of continuity have distinctly different characters. His middle definition is internally consistent and mathematically useful, but disallows the infinitesimals which he wished to use in place of limit theory. His final definition is consistent with the existence of infinitesimals, and consistent with his intuitions of continuity as an idealized smoothness, but it is not mathematically useful. Perhaps the most important lesson for us in all this is that Peirce's intuitions led him to be severely dissatisfied with Cantor's

---

[69]See Peirce [1976, v. 3, p. 127].
[70]Ibid., p. 127.
[71]See Dauben [1977, p. 131].
[72]Ibid., p. 131.

and Dedekind's theories of the continuum. Paul du Bois-Reymond expressed a similar dissatisfaction. Peirce, however, did not provide us with a system capable of satisfying both his intuitions and the needs of mathematics. In Chapter 8, I shall attempt to characterize the root of this dissatisfaction.

CHAPTER 7

Infinitesimal Interlude

This chapter will not be as short as the title seems to suggest; however, it will not be lengthy. Its purpose is limited: to defend the mathematical usefulness of infinitesimal quantities. As we have already noted, many have claimed that even if infinitesimals were consistently definable, given their non-additive nature (an infinitesimal plus a finite number equals that same finite number), such entities are mathematically useless. We saw in Chapter 4 how Cantor, in the midst of making this charge against infinitesimals, also preached patience (though with respect to his own transfinite quantities, not with respect to infinitesimals), attempting to persuade people to refrain from judging a new mathematical concept as illegitimate simply because it does not seem to have an immediate use. In this chapter, I would like to not only urge patience, but go beyond, arguing that some mathematical concepts may be best understood with the use of infinitesimal quantities. I will not prove their existence, their status as numbers, or even the consistency of their definition; I simply argue that if we can consistently define them, there are mathematical concepts to which they apply quite well.

In this chapter, I wish to present two examples of cases where the use of infinitesimals preserves a meaningful difference that limit theory overlooks, one example geometrical, and one from probability theory. Next, I wish to briefly indicate how Abraham Robinson creates a conservative extension of the real numbers, which yields a non-standard system of numbers including both infinitesimals and transfinites, as some indication of what such a system would look like. Finally, I end with a sketch of a proof by Euler in which he uses infinitesimals to great effect.

Our first example is drawn from geometry.[1] Imagine a circle with a line drawn tangent to it, like so:

FIGURE 7.1. Circle with Tangent

By definition of "tangent," the line touches the circle at precisely one point. Now imagine that we wished to measure the angle formed by the tangent and the arc of the circle. We can approach a measurement by fitting a rectilinear angle around our curvaceous one, like so:

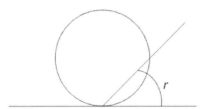

FIGURE 7.2. Curvilinear Approximation, Step 1

Let us call the rectilinear angle r, as above. It is clearly much bigger than the curving angle, but were we to make it smaller, the curved angle would be still smaller:

---

[1]This example appears in Waismann [2003, p. 221-222].

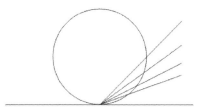

FIGURE 7.3. Curvilinear Approximation, Steps 2 through n.

And so on. In fact, no matter how small our r angle gets, the angle of the curve will always be smaller. The limit of the ever-decreasing rectilinear angles as we approach the curvilinear angle is zero; using limit theory to measure this angle would give us 'zero' as the answer. Yet the curved angle is clearly not zero, as that would imply that it was identical with the line itself, and the arc of the circle does make an angle distinct from the line. It is simply an angle which is smaller than any finite positive rectilinear angle, no matter how small – and thus, the angle is infinitesimally small.[2] Other circular and circular-to-linear angles can be constructed whose measurements are infinitesimal quantities, and in fact, the study of circular angles possibly depends on infinitesimals. Thus, if one wishes to numerically express the difference between a straight line and a curved angle, an infinitesimal is a helpful means of doing so.

The second example is drawn from probability theory.[3] Imagine someone tossing a quarter. After one toss, the odds that the quarter will come up heads are $1/2$. After two tosses, the odds of the coin coming up heads both times is $1/4$. We can generalize this progression: for n tosses, the odds that every single toss will be heads is $1/2^n$. Imagine that a coin could be tossed infinitely many times – let's say that God decides to while away some time with a quarter. What are the odds that God would turn up heads every single time – that is, infinitely many times out of infinitely many tosses? Following the formula above, the

---

[2]Waismann not only used this proof to demonstrate what an infinitesimal might look like, but also to *define* an "actual infinitesimal quantity" η, in the following manner: η is defined "as the angle which the circle of radius 1 forms with its own tangent" (ibid., p. 222) Despite the geometrical argument, the definition of an "actual-infinitesimal," and the insistence that one could construct a geometry that lacks the Archimedean Principle, Waismann did not believe there was a future for infinitesimals. In fact, he suggested that dealing with infinitesimal systems "is nothing but idle play." Ibid., p. 224.

[3]This example is taken from Charles Dodgson's infinitesimal theory. See Abeles [2000].

odds would be $1/2^\infty$, viz., one over two to the power infinity. As two to infinity results in an infinite number, the odds are one over an infinite number – which is an infinitesimal.[4]

Some might say that the chances of infinitely many coin tosses turning up heads every time is not an infinitesimal, but rather it is an impossibility, because in infinity, every possibility is played out, and clearly the quarter turning up tails is a possibility which must be actualized at some point. This is a mistaken view of infinity.

Though some philosophers, such as Hegel and Spinoza, have argued that with regards to ultimate infinity, every possibility must with necessity be played out at some point, there are many infinite collections in which this ultimate plenum does not hold. For example, the positive integers are infinite, and yet they do not include every possibility. They exclude fractions, for one. The integers are not an all-inclusive sort of infinity, and neither is an infinite coin toss. Even if we limit the infinite coin toss to relevant possibilities, excluding the possibility that it turns up pink elephants sometimes, it is impossible for all relevant possibilities to be played out. For example, both the scenario where the coin turns up tails every time but one and the scenario where the coin turns up heads every time but one are possibilities; it is logically impossible for them *both* to occur. Even in an infinite series, not every possibility can be expressed.

If we determine the odds using limit theory rather than calculating the odds as $1/2^\infty$ as Charles Dodgson does, our equation becomes

$$\lim_{n \to \infty} \frac{1}{2^n}.$$

As n grows, the fraction shrinks, and approaches the limit 0. Thus, limit theory calculates the odds of an infinite coin toss turning up heads every single time as zero, but since in probability theory, zero means "absolute impossibility," one should be careful before applying the limit. Though it is highly unlikely that a coin thus tossed would turn up heads every single time, there is no logical impossibility to the event. If in probability theory we wish to distinguish

---

[4]Charles Dodgson believed that this demonstration proves the existence of infinitesimals. Perhaps his argument is not as strong as he believed; but at the least, it shows one place where infinitesimal quantities could be usefully applied: in calculating infinitary probabilities.

between something being highly unlikely and something being impossible, preserving the formulation of the odds as $1/2^\infty$ and admitting infinitesimals into probability theory is one way to do so.

Further, in the infinite coin-toss example, every particular outcome has exactly the same odds, just as it does in a finite string of coin tosses. The probability that every coin toss would come up heads is an infinitesimal; the probability that the tosses would strictly alternate heads and tails throughout the series is an infinitesimal, and so on for any particular combination of heads and tails, as each solution to the problem is one possibility of infinitely many possible solutions. Thus, we must admit that either an infinite coin toss is logically impossible, or that each particular solution is equally likely – infinitesimally possible. It may seem tempting at this juncture to abandon the possibility of an infinitely long series of coin tosses; after all, it is a physical impossibility for any human to take up (and unlikely that God would toss a quarter through all infinity). However, we may wish to retain the ability to calculate odds in an infinite system, should we wish to view mathematics and the universe itself as potentially containing infinite sets; thus, abandoning the ability to mathematically distinguish between infinitely improbable and logically impossible should not be undertaken unreflectively.

The most developed non-Archimedean system is that of Abraham Robinson's nonstandard analysis, and I wish here to present an outline of how he creates the set of numbers he uses in his analysis. I will follow Robinson's own exposition.[5] He begins with standard first order logic, and proves the theorem

> Let K be a set of sentences in a language Λ [where Λ is a higher-order language than first order logic, and thus our relationship symbol can range over sets as well as individuals]. Suppose that every finite subset of K is consistent. Then K is consistent.[6]

Robinson uses this theorem to define the enlargement of K in such a way that if K is consistent then its enlargement is consistent as well. Furthermore, such a conservative extension retains many of the properties of the original set; thus,

---

[5]See Robinson [1996].
[6]Ibid., p. 27.

136   7. INFINITESIMAL INTERLUDE

the nonstandard enlargement of the natural numbers *N has many properties in common with the natural numbers themselves:

i) Every mathematical notion which is meaningful for the system of natural numbers is meaningful also for *N. In particular, addition, multiplication, and order are defined for *N.
ii) Every mathematical statement which is meaningful and true for the system of natural numbers is meaningful and true also for *N: provided that we interpret any reference to entities of any given type, e.g., sets, or relations, or functions, in *N not in terms of the totality of entities of that type, but in terms of a certain subset.[7]

Thus Robinson is able to create a larger mathematical system while preserving the functions necessary to mathematics. It is unfortunately beyond the scope of this book to investigate the philosophical or mathematical definition of continuity implied by Robinson's non-standard analysis, however, one imagines that a first step would be to show that the conservative extension of the real numbers necessary for calculus does not imply Dedekind continuity, i.e., that not every cut in Robinson's non-standard system is determined by a unique real number. This seems intuitively correct, since not all of the elements in Robinson's universe are in fact real numbers, as they are in Dedekind's mathematical system. While Robinson's non-standard analysis is strictly outside the scope of this book, it is worth noting that his thoroughly developed calculus has overcome many of the technical and philosophical problems which plagued the infinitesimal systems of the nineteenth century, drawing on them for inspiration and pushing the project through to a workable system. Work on this system continues to the present day, and mathematicians are beginning to find a variety of uses for a calculus which includes both infinitesimal and transfinite quantities.

In closing, it is worthwhile to exemplify the potential usefulness of infinitesimals in calculation by alluding to a proof of Leonhard Euler (1707–1783). Euler suggested a system of infinitesimals in his book, *Introduction to Analysis of the Infinite*. In chapter VII, Euler attempted to obtain infinite series expansions for exponential and logarithmic functions without using differentiation or integration. In doing so, he defines ω as an "infinitely small number, or a fraction

---
[7] Ibid., p. 49.

7. INFINITESIMAL INTERLUDE

so small that, although not equal to zero, still $a^\omega = 1 + \psi$, where ψ is also an infinitely small number."[8] He let $\psi = k\omega$, so that $a^\omega = 1 + k\omega$. He then presented an example of this function at work using a finite ω:

> He let $a = 10$ and $\omega = 0.000001$, so that $10^{0.000001} = 1 + k(0.000001)$. It follows (from a table of logarithms) that $k = 2.3026$. On the other hand, for $a = 5$ and $\omega = 0.000001$, he found that $k = 1.60944$. "We see," concluded Euler, "that k is a finite number that depends on the value of the base a."[9]

Euler goes on to expand a^x for a finite x by using the transfinite number j, defined as $j = x/\omega$. As such, Euler completely bypasses the need to refer to limits of any sort, instead operating with defined infinitesimal and infinitely large variables. Such a proof is ingenious, but of course controversial. In sketching the proof in his book, *Euler: the Master of us All*, Dunham hails the cleverness of the logarithm proofs, but simultaneously slanders the use of infinitesimals.

> Such reasoning hails from a pre-rigorous era. This is not to say, however, that it should be casually dismissed. On the contrary, it accurately reflects the standards of its day and, in that context, is both clever and compelling.[10]

By excusing the use of infinitesimals as "reflecting the standards of the day," Dunham echoes the many people who excuse racism and sexism of times past by calling the perpetrators products of their time. Yet many today do not see infinitesimals as shameful or embarrassing mistakes made by our ancestors who were unfortunately born in an unenlightened era; some view infinitesimals as useful tools in calculation, adequate and sometimes preferable substitutes for limit theory.

The preceding examples do not prove the existence or consistency of infinitesimals; they do not even prove beyond a shadow of a doubt that infinitesimals are mathematically useful. What they do accomplish, however, is to show

---

[8]See Euler [1988, p. 92].
[9]See Dunham [1999, p. 24–25].
[10]Ibid., p. 29.

that infinitesimals are not *prima facie* useless simply because they are non-additive, that there may be some situations in which infinitesimals work just as well as, or perhaps even better than, limit theory.

We must also keep in mind that mathematics frequently proceeds without deciding beforehand what mathematical systems will be of use to later generations. While Newton developed the calculus with uses firmly in mind, Leibniz did not. One suspects that analyzing the square root of negative numbers was not first done in order to find an application for imaginary numbers, but rather in order to see what would happen if one analyzed imaginary numbers.[11] As Dunham puts it:

> Mathematics, of course, has been spectacularly successful in such applied undertakings [as determining planetary orbits and balancing checkbooks]. But it was not its worldly utility that led Euclid or Archimedes or Georg Cantor to devote so much of their energy and genius to mathematics. These individuals did not feel compelled to justify their work with utilitarian applications any more than Shakespeare had to apologize for writing love sonnets instead of cookbooks or Van Gogh had to apologize for painting canvases instead of billboards.[12]

While "uselessness" may be a cutting charge in feminist political theory or in sociology, mathematics has a habit of discovering mathematical entities first, and finding uses for these theories later, if at all.

The next chapter is dedicated to an analysis of continuity itself. However, the analysis presented will suggest a natural place for a system of mathematics which contains infinitesimal quantities.

---

[11]Cardan developed imaginary quantities as a way of dealing with equations involving the root of negative numbers, but he thought that actually solving such equations was impossible. See Smith [1929, p. 201–202].

[12]See Dunham [1991, p. vi].

CHAPTER 8

Conclusions

8.1. Introduction

We have seen four different theories of continuity; here we shall compare them directly, and draw some philosophical conclusions about continuity in general and mathematical continuity in particular. In this chapter I wish to define the concept of "compositional continuity," discuss the viability of compositional theories, and apply this analysis to each of our four figures. I will conclude that, viewed in this manner, Cantor and Dedekind present flawed theories of continuity, and that, despite his criticisms of Cantorian continuity, so does Peirce. The most promising theory of the four is that of du Bois-Reymond's Idealist. The analysis of continuity I will present here, together with the Chapter 7 defense of infinitesimals as mathematically useful, will make apparent a logical place for infinitesimals in relation to theories of continuity.

This chapter will thus consist of four main parts. First, drawing on the work in Chapters 3 through 6, Section 8.2 will summarize the key philosophical positions of each of our four figures on continuity and infinitesimals. Next, in Section 8.3, I shall define "compositional continuity," expand on Chapter 2's exposition of Aristotelian continuity, and address Aristotle's argument against composing continua from points. From there the remainder of the chapter is philosophical analysis. Section 8.4 will show that Cantor and Dedekind both have compositional theories of continuity, and both suffer from certain philosophical problems as a result. Section 8.5 will discuss the complicated way in which Peirce, despite his criticisms of compositional continua, fell into a certain type of compositionality himself, and Section 8.6 will address and evaluate du Bois-Reymond's theory of continuity, and discuss the role of infinitesimals in continuity itself. Finally, in Section 8.7, I will end with a few concluding remarks.

8.2. Summary of Our Four Figures

In the following summaries, I pay particular attention to each man's theory of continuity, his particular view on infinitesimals, and the ways in which continua and infinitesimals interact in each theory.

8.2.1. Richard Dedekind. As we saw in Chapter 3, the impetus behind Dedekind's theory of real numbers and theory of continuity was his dissatisfaction with the geometrical nature of the calculus. He wished to create a truly arithmetized calculus, one that did not rely upon geometrical intuitions about curves and tangents. His theory of real numbers and continua reflects this essential dissatisfaction; indeed, Dedekind gave a purely arithmetic interpretation of these mathematical concepts. Through his general theory of number, combined with his real number theory, Dedekind wished to construct a plausible logical progression, beginning with counting finite, discrete objects, and gradually adding numerical complexity as needed for completeness and calculation, ultimately arriving at the set of all real numbers and the ability to quantify complex phenomenona such as motion at an instant. The logical progression is of course different in significant ways from the historical progression of the creation of the calculus, but he wished to build the progression such that it took no detours at all through geometry or intuition, but was firmly based in sets of numbers from beginning to end. Finite sets of counting numbers give rise to the function of succession, and the naturals are created; the naturals, plus functions such as subtraction and division (which naturally arise from addition and multiplication) create the set of rational numbers. The irrational numbers are created from the rationals, through Dedekind cuts, thus producing the set of real numbers. Dedekind argued that this set exhibits the essence of continuity, and therefore no more was needed to complete measurements and calculations involving continuous objects.

The real numbers, Dedekind argued, exhibit the same essence of continuity as the geometrical line, and therefore, can replace geometrical intuitions entirely in the calculus. Recall from 3.2, the essence of continuity for Dedekind is thus.

8.2. SUMMARY OF OUR FOUR FIGURES 141

> If all points of the straight line fall into two classes such that every point of the first class lies to the left of every point of the second class, then there exists one and only one point which produces this division of all points into two classes, this severing of the straight line into two portions.[1]

The ability to cut a continuous entity anywhere, and have the division necessarily fall at an element of the entity, and not between entities, defines continuity for Dedekind. For the straight line, the elements in question are points; for the set of real numbers, the elements in question are numbers. While Dedekind does not directly address the issue of whether the line is composed of points, or if the line contains nothing but points,[2] he is quite clear that the real number continuum is composed of numbers and contains only numbers. Further, he is able to prove that this property, this essence of continuity, holds for the real numbers as created by Dedekind cuts. That is, once the full set of real numbers is created, there *must be* a number wherever you cut the set.

Of course, due to Dedekind's definition of irrational numbers, this essence of continuity follows for the reals almost trivially. An irrational is created when one finds a cut in the rationals which does not occur at a number; having found such a cut, we create a number at its location, and term it an irrational. Thus, it follows easily from this definition that wherever there is a cut on the reals, the cut happens at a number. Dedekind's principle of continuity clearly holds of the reals created in this manner. Also, it follows that this real number continuum contains nothing other than the real numbers themselves; the logical progression from finite counting numbers to the full set of reals assures us that nothing was added at any stage except numbers. Further, nothing more *can* be added, since wherever a gap in the system occurs, an irrational number is created to fill the gap; there is no opportunity for non-real numbers to be included in such a

---

[1] See Dedekind [1963, p. 11].

[2] We can infer, however, from this essence of continuity, that the line already contained all of its points; he thus would have rejected a characterization of the geometrical line such that we create a point on the line through the act of division itself. When we divide a line, either a point exists at this division or it does not; if we are assured that a point exists at every place of division, then we can call the line continuous. Interestingly, the parallel to Dedekind's real numbers does not hold; the process of the creation of the real numbers requires that, should our cut fall between rationals, we create a number through the act of division itself.

composition. Intuitively, this is why Dedekind's real number system implies the Archimedean Principle, and is thus incompatible with infinitesimal quantities.[3]

In sum, Dedekind's real number system is built from the foundation of the counting numbers, and is itself complete enough to satisfy our principle of continuity. The system itself contains nothing but rational and irrational numbers; it does not contain segments or intervals as metaphysical atoms (the way du Bois-Reymond's Idealist system does), and is mathematically incompatible with infinitesimal quantities at its heart. While Dedekind did not attempt to prove that the points on the geometrical straight line are in any sort of correspondence with the numbers in the real number system, he did, at least, create a number system which shares the same principle of continuity found in the straight line, and, unlike the rational numbers alone, can be argued to be as rich in "number-individuals" as the straight line is in "point-individuals."[4]

8.2.2. Georg Cantor. Cantor's theory of numerical continuity is similar to Dedekind's in many ways. He, too, built a real number system based on rational numbers, and he, too, believed that calculus could be founded upon purely arithmetical principles. He was vehemently opposed to the idea of infinitesimal magnitudes; however, his system of numbers was much more explicitly open to the inclusion of non-real numbers than was Dedekind's. For example, he firmly believed that imaginary and complex numbers were an important part of our mathematical systems; and, of course, he wished to extend our concept of number to include the transfinite numbers, which are not a part of the real number

---

[3]We can, however, continue our logical progression *beyond* the set of reals, as Cantor did when he created his transfinites. Once we have, through Dedekind's logical progression, created the naturals and the reals, we have demonstrably created infinite sets of differing magnitudes, thus, the possibility of appending cardinalities to the differing sets becomes not only possible, but quite appropriate. Cantor's creation of infinitely many such cardinals based on sets of the first two transfinites follows Dedekind's method of manipulating the numbers we have with functions naturally arising from our systems, and creating numbers to satisfy these operations for the sake of completeness. Infinitesimals can thus be viewed as the numbers necessary to satisfy completeness of operations for transfinite numbers, plus the already well established operation of taking an inverse of a number. Thus, infinitesimals are not, strictly speaking, contradictory with Dedekind's system. If the real numbers do represent true continuity, then infinitesimals cannot logically be a part of that continuum, but their existence outside of the continuum is not contradictory, but rather, follows Dedekind-approved number-creation tactics.

[4]See Dedekind [1963, p. 9].

system (though an important tool, Cantor felt, to analyzing and understanding this system). Though an advocate of the use of non-real numbers of varying magnitude, he still maintained that a continuum can be constructed from real numbers alone, and that infinitesimals not only have no place in a continuum, but are self-contradictory and hence have no place in our mathematics.

Though Cantor's irrational numbers are, like Dedekind's, built from sets of rational numbers, he does not have quite the same foundational approach as Dedekind. His chief anxiety was not to eliminate references to geometry, but rather, to make mathematics more elegant, more functional, and better grounded. Thus, while Cantor's real number theory, definition of continuity, and attitude toward infinitesimals are all similar to those of Dedekind, his philosophical approach is markedly different.

Cantor's primary philosophical concern in creating irrational numbers was to avoid a circular definition; he believed many, if not all, mathematicians before Weierstrass assumed the existence of particular limits in creating their irrational numbers, and then went on to define irrational numbers in terms of these limits. Thus, Cantor's sets of rational numbers – Cauchy sequences, or, as Cantor called them, fundamental sequences – were *associated* with particular symbols, but Cantor was careful not to assume these symbols were the limits of their associated sets, and even more careful not to assume that these symbols were, in fact, numbers. Rather, he provided evidence for the limit-like nature of these associated symbols, and went to great lengths to prove that mathematical functions such as addition, subtraction, and equality, held for these symbols before he allowed himself to refer to them as magnitudes.

Cantor's definition of continuity was just as rigorously constructed. Rather than intuiting the essence of continuity and then arguing that his real numbers lived up to our expectations, he instead provided necessary and sufficient conditions for the continuity of any set, and demonstrated that his real numbers met these conditions. Rather than arguing that his mathematical continuity was consistent with other continua, such as that of space, time, or the geometrical straight line, he argued that his mathematical definition was logically prior to any other intuition of continuity, and that we should understand the continuity of space and time in terms of his necessary and sufficient conditions, rather than the reverse. Thus, these point continua of Cantor's provided the true essence of continuity, not by developing our intuitions, but by grounding them in rigorous theory.

Despite these differences, and Cantor's openness to non-real numbers such as the transfinites, his theory of continuity is still firmly committed to the idea that there are no logical atoms in a continuum other than the point-elements – such as numbers, in the case of mathematical continua, and points, in the case of geometrical continua. He referred to continua as "point continua," and believed that it is sufficient to show one continua reducible to another to show that the elements (points or numbers as the case may be) are in one-to-one correspondence. Thus, the essence of continuity, for Cantor, is contained entirely in his two necessary and sufficient conditions for point continua – connectedness and perfection of sets. The mathematical definitions of these terms were thoroughly explained in Chapter 4. In sum, a set exemplifies perfection if and only if every derived set is contained in the set itself (thus, the collection of limit points of a set does not extend beyond the set itself). The connection of a set can be variously interpreted (see, for example, Peirce's attempts to analyze Cantor's connectedness), but at the very least, it implies everywhere denseness of the set.

As for infinitesimals, Cantor argued in the *Grundlagen* that even if infinitesimal magnitudes could be consistently defined, they were useless (though he argued in the same work that pure mathematicians should not concern themselves with the usefulness of their mathematical creations). Later, however, he argued that they could not be consistently defined, that infinitesimals were inconsistent with the very notion of linear magnitude, and thus, an infinitesimal magnitude was a contradictory concept. The argument Cantor put forward to try to show infinitesimals contradictory relied heavily upon the Archimedean Principle as the main criteria for an understanding of the concept of magnitude, though his own beloved transfinite numbers also fail to meet this criteria.

8.2.3. Paul du Bois-Reymond. Like Cantor, du Bois-Reymond was led to his interest in the philosophical underpinnings of infinity and continuity through his sometimes controversial mathematical systems. In du Bois-Reymond's case, his work with infinitary functions (his Infinitärcalcül) raised foundational concerns; by using the limit operation to organize functions with infinite ranges and domains, his system brought into sharp relief questions

8.2. SUMMARY OF OUR FOUR FIGURES

about infinity, as well as questions about the nature of limits themselves. Like Dedekind, du Bois-Reymond was openly dissatisfied with non-rigorous approaches to these difficult concepts found in pedagogical situations, and du Bois-Reymond's dissatisfaction led him to attempt a systematic development of the intuitive underpinnings of our mathematical concepts.

He found two intuitions at the ground level, and claimed each of us, if we thought carefully enough about mathematics, would find both of these intuitions equally compelling. First was the Empiricist intuition, by which we feel that mathematics must spring from our experience of the world. According to our Empiricist intuitions, counting must spring from our need to count objects, number derives from the notion of counting, and all mathematical concepts could and should be either traced back or reduced to our experience of the world around us. Second was the Idealist intuition, from which we get our sense that mathematics can be developed as far as logical organization can take it, regardless of whether it retains its tether to the empirical world. He believed that these intuitions were not only competitors, but in open conflict, with virtually no shared ground between them in metaphysics, application, or the acceptance or non-acceptance of mathematical objects into our canon.

These competing intuitions gave rise to different philosophical attitudes toward mathematics, which, in their turn, gave rise to differing views on what mathematical entities are acceptable, and what mathematical processes should be followed. Thus, a mathematics which respected our Empiricist intuitions would proceed slowly, checking with empirical reality at each step of its development, whereas a mathematical system developed along Idealist lines would proceed more rapidly, avoiding contradiction or chaos but adding any other elements that seemed helpful and logically satisfactory. Crucially for us, an Idealist mathematics contains geometrically idealized objects and procedures; it also contains transfinites and infinitesimals. An Empiricist mathematics, on the other hand, forbids objects so idealized that they can no longer be said to be drawn from experience. Extensionless points and one-dimensional lines are impossibilities, and obviously, so are actually infinitely large or small magnitudes.

Thus, according to du Bois-Reymond, the Empiricist intuition in mathematics leads to the conclusion that there is no such thing as mathematical continuity. Perhaps there is continuity in the physical world, but if so, it is beyond the direct experience of human beings, and thus, our mathematics and geometry contain no such thing. The geometrical line is, for the Empiricist, as long as we want, but not actually infinitely long; as thin as we want, but not actually only one dimension; as divisible into as many parts as we want, but not actually infinitely divisible. The Empiricist may believe that we can find a point wherever we wish one on the geometrical line, but that claiming actual continuity of the line is entirely different, and unwarranted. What holds of the geometrical line, according to Empiricist intuitions, holds even more strongly for the number line. The numbers are not infinitely dense, nor are there infinitely many, and certainly they do not form a continuous set. However, du Bois-Reymond's Empiricist believes, with Aristotle, that "as many as we wish" is quite enough for mathematics to function well.

A mathematics built on Idealist intuitions, on the other hand, contains continua. Idealist mathematics has no problem with the infinitely large, infinitely small, or infinitely smooth. The geometrical line is continuous. A continuous set of numbers, in Idealist mathematics, is infinite, dense, and has no gaps, but it is not built from the ground up – it is not built from empirically established phenomena. Even the Idealist admits that continuity is quite beyond our experience.[5]

Further, the Idealist continuum necessarily contains infinitesimal quantities. The belief that the geometrical line is infinitely divisible leads, du Bois-Reymond argued, to the conclusion that the geometrical line contains more than points; it also must contain infinitesimally small intervals. If the number line is to measure the geometrical line, and other infinitely divisible and continuous entities, it must therefore have an infinitesimal quantity with which we can measure these infinitesimal segments. A thorough understanding of how these infinitesimal quantities interact with each other and with finite quantities is necessary (and thus is developed by du Bois-Reymond).

Du Bois-Reymond believed that the internal struggle between our Empiricist and Idealist intuitions was an insoluble one, one that we must continue to

---

[5]Du Bois-Reymond's insistence that we do not experience continuity is not uncontroversial; many, including Peirce, believe that we experience continuity in our daily lives, and, as was noted above, Kant believed that continuity was a necessary condition of any of our experiences. It is unclear how the Empiricist's mathematics would change, had du Bois-Reymond believed that continuity was part of human experience.

grapple with as we create new mathematical systems and analyze the philosophical foundations of our current ones. It is quite notable, however, that both the mathematics developed strictly according to Empirical intuitions and that developed according to Idealist intuitions are radically different from the mathematical systems of Dedekind and Cantor.

8.2.4. Charles Sanders Peirce. Peirce believed that continuity was one of the key concepts to understanding the nature of the world, that it was important to sciences as diverse as botany, history, and psychology. He defined continuity early in his career as, "the passage from one form to another by insensible degrees,"[6] but found this definition too imprecise and sought to formalize it. This imprecise first attempt to pin down the concept of continuity is instructive, however, as it reveals his early intuitions on the subject. He would spend much time and ink over the course of his career attempting to find a precise philosophical and mathematical understanding of continuity that lived up to his intuitions on the subject. He first tried to formalize this intuition by defining continuity simply as density, but soon judged this inadequate, and went on to define a theory of continuity that much resembled those of Dedekind and Cantor; in fact, he was explicitly influenced by Cantor in his thinking on continuity, as he was in many of his mathematical views.

This Cantorian definition of continuity contained three conditions, individually necessary and jointly sufficient. A continuous set, according to Peirce at this stage of his thought, must have the properties of (i) infinity and nondenumerability, (ii) density, and finally the set must (c) contain all of its limit points. In (c) we recognize a concept quite similar to the perfect sets of Cantor. This definition shares many properties with Cantor's definition; in addition to the property of perfection, Peirce's continuity is meant to hold of sets consisting of individual members, such as numbers or points, or perhaps instants.

Though Peirce accepted and defended this mathematical continuity, which is similar enough to Dedekind and Cantor's that we can derive Dedekind's from Peirce's, Peirce believed at the time that infinitesimals were an important part of any continuum. Whether Peirce saw this contradiction (recall that Dedekind-continuity implies the Archimedean Principle and thus is inconsistent with the existence of infinitesimals) is not clear; what is clear is that he soon became

---
[6]CP 2.646.

dissatisfied with this definition as well. While his first attempt at a definition characterized his intuitions, it lacked precision; this revised definition, although quite precise in the sense that it allowed one definitively to determine whether any given set was or was not continuous, no longer satisfied Peirce's intuitions.

His third and final definition of continuity was an attempt to characterize his intuitions with greater precision. Under this final definition, Peirce abandoned density as a requirement, replacing it instead with the requirement that, for an entity to be continuous, "all of its parts must have parts of the same kind."[7] This, he judged, was the essence of continuity – a precise characterization of the "insensible degrees" intuition he expressed so much earlier – and this essence of continuity had vast ramifications for the nature of a continuum. He took as his inspiration for this new definition a deeper understanding of Immanuel Kant's definition – that a continuum is such that every part of it resembles every other part, in every respect except for magnitude itself. Thus, no longer could Peirce accept the idea that continuity might hold of sets of individuals such as points, for a point would be a part of that continuum that differed markedly from other parts, such as segments.

This sort of continuum could be infinitely divided, and every division would produce a part which was isomorphically identical to every other part. Continua could not be composed of points, as that would make points essential building blocks, essential parts, which were themselves non-continuous. Nor could a continuum be decomposed into points, for similar reasons However, in so far as these continuous entities could be measured, one could create points in particular places to aid our measurement, but wherever we place a point for purposes of measurement, we disrupt the continuity of the entity we are measuring.

This final definition of continuity also featured infinitesimals in an important role. Infinitesimals, he argued, could themselves be infinitely divisible. Thus, they could be used as points of measurement without essentially disrupting the continuity of the object to be measured; they were precise enough to determine a location, and yet were themselves continuous magnitudes, identical to a finite segment in every respect except for size. Thus, for Peirce, the elements of a continuum varied depending on what we did with the continuum.

---

[7] CP 6.168.

All continua were divisible, with each division producing continuous parts; one could find points on a continuum, at the cost of disrupting continuity, or one could instead use infinitesimals as a continuity-preserving method of measurement

However, Peirce did believe that continuity thus defined could be built out of discrete elements. One simply continues to add discrete elements until their discreteness disappears. Thus, in the case of the numbers, Peirce believed that if one took every number in a set and replaced it with all of the real numbers between zero and one, and then took that set and replaced all of those numbers with the real numbers between zero and one, etc., eventually the numbers would lose their distinct nature and 'weld' together, forming a continuous mass of undistinguished elements. Peirce himself believed this "supermultitudinous" collection completely satisfied our intuitions about the nature of continuity; however, he also believed that the collection was too unwieldy to be useful to the science of mathematics.

8.3. Compositional Continua and Aristotle

Having summarized the four figures addressed in earlier chapters and their approaches to continuity, we are now able to analyze the various definitions presented in terms of their compositional nature. Thus, in this section, I would like to accomplish three things. First, I would like to define precisely what I mean by "compositional continuity." Second, I would like to analyze Aristotle's argument against composing continua from indivisibles, and show why it is not applicable to continuity composed from points or numbers. Third, I wish to argue that despite the failure of Aristotle's argument, Aristotle's analysis does succeed in capturing our intuition of continuity itself, and thus, definitions which are compositional in nature are at best problematic. In the sections which follow, I will argue that three of our four figures adopt compositional definitions of continuity, and thus fall prey to these problems.

By "compositional continuity" I simply mean a theory of continuity which necessitates or allows a continuum to be composed of non-continuous elements. To draw on a finite and non-continuous analogy, a set of five apples is composed of five discrete apples, and thus is compositional; an apple pie, before it is sliced, cannot be said to be composed of five slices of pie (though in the case of finite entities, once we create the divisions, we can then recompose the entity out of those divided entities). An entity containing infinitely many elements can be said to be composed of those elements if (i) the elements contained within the entity are logically prior to the entity as a whole, and (ii) the entity contains nothing else which is similarly logically prior to the entity as a whole. By "element" is understood any entity, mathematical, geometrical, or otherwise, which is not itself a continuous entity. By this definition, one can construct a continuous segment of a line from shorter segments of a line, each of which is itself continuous, without viewing the larger segment as thus compositional in nature. If, however, the segment is viewed as composed from infinitely many points, as points are not themselves continuous, this would be a compositional continuity.

As was discussed in Chapter 2, Aristotle believed a continuum could not be composed from indivisible elements. His explicit argument against compositional continua does not hold against a real number continua even should it be compositional; but it is in his general definition of continuity that we find the intuitive properties of continuity so well expressed, and thus, it is to Aristotle we turn for inspiration expressing what is troublesome about a point continuum as defended by Cantor and Dedekind.

Recall, the essence of Aristotle's argument against compositional continuity relies on a reductio: if we assume the existence of a continuum composed of elements, it immediately follows that two points must be next to each other. An exhaustive examination of the concept of 'next to' reveals that no two indivisible elements can be next to each other, for any type of 'next to.' This argument fails to apply to real number continua, or modern conceptions of point continua that include a line composed of points, as a minimum requirement in modern notions of point continua, as we saw expressed in Cantor, Dedekind, and Peirce, above, is density – between any two members of the set, there must be another member, and in fact, there must be infinitely many. Thus, in such sets of numbers or points, compositionality does not immediately imply that two elements are next to each other; sets in which this is possible are explicitly excluded from consideration as continuous sets.

Compositional continuities which include density as an explicit feature do, however, violate Aristotle's essence of continuity, his own basic definition of the term. Recall that he defined the continuous as "that which is divisible into divisibles that are always divisible."[8] This definition includes infinite divisibility, of course, but also has at least one feature of what we termed above the "mirror quality" of continuity, expressed in Kant, Peirce, and du Bois-Reymond. As Peirce put it, our most basic intuition about the nature of continuity is that "all of [its] parts have parts of the same kind."[9] The intuition that all essential parts of a continuum must themselves be continuous is one which is persistent throughout literature on continuity, and is echoed by Kant, when he claims that, "the property of magnitudes by which no part of them is the smallest possible, that is, by which no part is simple, is called their continuity."[10]

Compositional continua violate the mirror quality. By composing a continuity from non-continuous elements, those non-continuous elements are not only essential parts of the continuum, they are logically prior to the continuum itself. And yet, they violate the mirror quality by lacking continuity. Non-continuous elements, such as points, are not infinitely divisible, are not isomorphically similar to every other part of a continuity, and, at least in compositional views of a geometrical line, violate Kant's definition by requiring the point to be the smallest possible element; the point can, in these cases, be viewed as "simple" in precisely the way in which Kant wishes to forbid it.

Du Bois-Reymond offered a related criticism against compositional continua when he wrote

> Points are devoid of size, and hence no matter how dense a series of points may be, it can never become an interval, which always must be regarded as the sum of intervals between points.[11]

Thus, while Peirce focuses on the difficulty of non-continuous parts forming a continuum, du Bois-Reymond emphasizes the inadequacy of points themselves as building blocks of continua. Philip Ehrlich adds that du Bois-Reymond "was not alone among late 19th-century thinkers in believing that, if a continuous line is to be regarded as composed of elements, these elements must themselves

---

[8]Aristotle's *Physics*, Book VI, 232b p. 23.
[9]Peirce, CP 6.168.
[10]Kant, *Critique of Pure Reason*, A 169 (B211).
[11]See du Bois-Reymond [1887, p. 66].

be extended."[12] Of course, these elements must themselves also be continuous, for if a continuum were to be formed of discontinuous extended elements, the result would be a discontinuous entity.

Aristotle's definition of continua as that which is divisible into parts which are themselves divisible fits with our intuitive sense that continua are not destroyed by mere division, and that continuity ensures a particular kind of sameness. Recall the examples from Chapter 1 of the division of tables and of water; physically dividing a table produces something not at all table-like, and while dividing a glass of water into two results in smaller amounts of water, it is possible to divide water enough times to break down the molecules and even atoms themselves, proving water to be non-continuous. In fact, we could say that water has a smallest element; the water molecule, H_2O, is the smallest element of water; further division creates something that is non-water. Compositional continua are anti-intuitive precisely because they violate our sense that continuity, at the very least, must guarantee that small parts of the continuum must resemble the larger parts in every manner but size. Small areas of space must resemble large areas of space; small units of time must flow in a manner similar to large units of time; if these things failed to be generally true, we would judge space and time as noncontinuous entities.

8.4. Continuity, Cantor, and Dedekind

Cantor clearly believes that the geometrical line is composed out of points, and Dedekind most likely believes so as well; to this extent, at least, they violate this intuitive mirror quality of continuity. However, the primary focus of this book is not geometrical but rather numerical continuity. Both Cantor and Dedekind clearly believe the real number continuum is composed of real numbers; whether they fall sway to the above criticism depends entirely on whether a real number can itself be considered a continuous or a non-continuous entity. If a real number can be considered a continuous element, the above criticism does not apply. It is not a straightforward question, given the rather unique nature of numbers themselves. Asking whether the number 2 is a continuous entity rather seems like a category error of sorts; such concepts as continuous

---

[12]See Ehrlich [1994, p. x].

8.4. CONTINUITY, CANTOR, AND DEDEKIND

or discrete do not easily seem to apply to individual numbers. Further, if we interpret a particular number as simply a magnitude, such as a particular stretch on a geometrical line or a particular length of space or time, one may suppose that a number could in fact be a continuous object. The number two, after all, at least meets the "infinitely divisible" criteria of continuity, as we can divide two by any other number, and divide that result again, and again, infinitely.

However, there is good reason to suppose that Cantor and Dedekind (and Peirce as well, as we shall see in the next section) believed numbers to be discrete entities, and thus their theories of a real number continuum would both create a compositional continuum and fall under the criticism of failing to preserve the intuitive mirror quality criteria of continuity. Cantor is fairly explicitly committed to considering individual numbers as discrete entities; his continuity is created from individual members of a set, which could be interchangeably points, numbers, or instants; numbers, for Cantor, are similar in nature to points, and thus, we can assume that he believed numbers, like points, are elemental, non-continuous elements. His explicit commitment to the Dedekind-Cantor axiom, which assumes that the set of real numbers and the points on a geometrical line correspond exactly is further proof that he takes the real numbers, not as the magnitudes which they measure, but as discrete points on the number line, individual locations blocking off magnitudes but not themselves representing continuous magnitudes.

Dedekind was not as explicitly committed to the Dedekind-Cantor axiom in his work, though he does seem to have it at least partially in mind in his construction of the real numbers and his definition of continuity. But recall, Dedekind's set of real numbers is built foundationally, beginning with the natural numbers, and adding numbers of different types as needed, until we reach the set of real numbers, which, he argues, satisfies our intuitions about continuity and is thus a continuous entity. The first level of construction, for Dedekind, is the ability to count finitely many discrete objects, and not the measure of length or width; thus, at the least, it seems as though Dedekind's most foundational numbers were either discrete in themselves or used to measure discrete objects; in either respect, they are non-continuous. Further, continuity, in Dedekind's mind, clearly comes from having enough of a proliferation of individual numbers to guard against the possibility of gaps. Were the number two viewed as a continuous magnitude, reflecting the distance between zero and two, then we would not have to add more numbers between zero and two to ensure the absence of gaps and thus finish the continuity. Dedekind's real number continuum, therefore, was similarly composed of non-continuous elements.

The problem with Cantor's and Dedekind's definitions of continuity is not that they contain non-continuous entities. A stretch of time may contain a discrete event; an area of space may contain a discrete entity. It is the definition of continuity as *logically dependent upon* the discrete that violates our intuitions. The difficultly comes only when continuity, which invokes an entity whose parts resemble each other no matter how small they become, is defined as necessarily, logically dependent upon entities which do not resemble the larger entity at all.

Defining continuity as *point continuity* as Cantor does – as logically composed of discrete elements – and then claiming that this is not only a good definition of continuity but the essence of continuity itself, is to make a category error. Indeed, Cantor goes further than to claim that a continuum can be formed from numbers whenever a set of numbers has the correct properties; he claims that this definition of continuity is necessary to comprehend all continua, even those such as space and time. If by *to comprehend* Cantor meant *to quantify and measure*, then he is correct. In order to analyze space, time, or any continua mathematically, we must have a tool of measurement that helps us understand continuity in discrete terms, as measurement is necessarily discrete. A line can only be measured if we posit points upon the line, and units of measurement that correspond to those points. However, if by *to comprehend* Cantor meant that to understand the deeper nature of the continuous, one must first understand the properties of the sets he defines as continuous, then he has conflated the properties of the tool used to measure continuity with the essential properties of continuity itself. Simply because discrete elements such as numbers and points are necessary to measure, quantify, or manipulate continuous entities does not mean that these continuous entities are *necessarily composed* of discrete elements.

Dedekind showed a similar conflation when he defined the essence of continuity with the geometrical line.

> If all points of the straight line fall into two classes such that every point of the first class lies to the left of every point of the

8.4. CONTINUITY, CANTOR, AND DEDEKIND

second class, then there exists one and only one point which produces this division of all points into two classes, this severing of the straight line into two portions.[13]

On the face of it, this seems a reasonable approach to continuity, and it certainly captures our intuition that continuity contains no gaps, or essential elements that are metaphysically different from the continuous entity itself. However, by focusing on this aspect of the geometrical line as the essence of continuity itself, Dedekind makes explicit the requirement that all continua contain an infinite plenum of discrete elements. Were we to attempt to apply this idea to time, for example, the first thing we would do is divide time – a normal enough occurrence. Suppose at precisely noon today I divide time into the past and the future; for any time in the past, it is clear it exists before any time in the future. The point of division, the present, is only useful as a tool of analysis if it is itself without time rather than an interval of time. So far, so good; time fits Dedekind's analysis in so far as it is divisible into two such classes, and the division happens at a point; neither at an interval of time, nor at a place somehow outside of time itself. However, if we apply Dedekind's essence of continuity more literally to time, it requires time to consist of infinitely many instants, and that the division between the past and the present consists of a division *of these instants*, one of which is singled out as the divisor for the others.

I wish to avoid veering too deeply into the philosophy of time, and thus wish to avoid pronouncing on the metaphysical status of instants, but it seems clear that the continuity of time does not itself require an infinite plenum of instants. It requires only that time take no breaks on its path, no gaps or jumps or other discontinuities. Instants are only required if we wish to analyze time, if we wish to understand time as the sort of thing we could measure, and separate into the past and the future. Thus, while one may wish to posit or even prove the existence of infinitely many instants in time, time as a plenum of instants does not follow from its continuity, even though division and analysis of time does sometimes require a unit of analysis which is itself non-continuous. This is what I mean by claiming that Dedekind conflates the means of measuring continuity with the features of continuity itself: in order to analyze and quantify a continuum such as time, non-continuous elements such as instants

---

[13]See Dedekind [1963, p. 11].

are desirable. However, our use of such elements in no way necessitates their *constitutive inclusion in the continuity itself.* All we must assure ourselves is that there are enough instants for us to measure time effectively, that is, that 'instant' is a useful tool for analyzing time.

Historically, proving that numbers were up to the task of measurement was a long-fought battle. Due to the incommensurability of the unit, it seemed as though numbers were a particularly bad tool of measurement when one wished to measure space; there were identifiable discrete points which could be found using the tools of geometry to which no number corresponded. The inclusion of irrationals in our numerical canon assures us that we can create a system of numbers of which Dedekind's essence of continuity holds – that is, that there are no longer identifiable gaps in the real number system, and each division of these numbers must occur at exactly one number. The correct conclusion to draw from this fact is not, however, that sets of numbers themselves are continuous, but rather that numbers so fashioned are in fact a good means of measuring space. Dedekind's system of real numbers makes the Cantor-Dedekind postulate feasible. We can postulate without contradiction or incommensurability the correspondence of points on a line with numbers in our real number system. However, points themselves are not an essential feature of the continuity of the line, but rather only an essential feature of our capacity to measure and quantify intervals of the line.

Thus, neither Cantor nor Dedekind produce an intuitively satisfying theory of continuity, and the primary reason behind this failing is the conflation of a property of a measurable entity with the tools for measuring and comprehending the entity which has that property. By defining continuity itself in terms of the points and numbers necessary to break continuity down into analyzable pieces, Cantor and Dedekind have, as Philip Ehrlich put it, reduced the continuous to the discrete.[14]

---

[14]See Ehrlich [1994, General Introduction, p. x.].

8.5. Continuity and Peirce

Both Peirce and du Bois-Reymond reject the Cantor-Dedekind style construction of continuity, and both criticize this type of construction in ways similar to the criticisms above. However, the alternative theories of continuity they present are substantially different from each other, and only one theory, that of du Bois-Reymond's Idealist, overcomes the criticisms raised against Cantor and Dedekind. In this section, I shall analyze Peirce's supermultitudinous continuity, and show how it retains some of the assumptions which led Cantor and Dedekind to trouble.

Recall, Peirce first created a theory of continuity that was quite similar to that of Cantor's and Dedekind's, and was in fact directly influenced by his study of Cantor, before rejecting it as not satisfying to Peirce's intuitions about continuity. He then based a new theory of continuity on his new understanding of Kant's definition of continuity as that property of magnitudes "by which no part of them is the smallest possible."[15] This is quite similar to the Aristotelian insight into continuity we quoted above, whereby Aristotle defined the continuous as that which can be divided into parts which are themselves divisible. Peirce took this intuition seriously, creating a theory of continuity which respected it.

Peirce's final theory of continuity, which tried to unify these disparate insights, does in fact present us with a continuum of which no part is the smallest possible, and which can always be divided into parts which are in turn infinitely divisible themselves. Supermultitudinous continuity does respect the mirror quality of our intuitions; it is precisely this feature Peirce had in mind when he formulated this theory. However, the resulting supermultitudinous collection, by beginning with a set of discrete objects such as points or numbers and constructing it from there, does not overcome du Bois-Reymond's criticism of compositional continua. Recall from above, du Bois-Reymond argued

> Points are devoid of size, and hence no matter how dense a series of points may be, it can never become an interval, which always must be regarded as the sum of intervals between points.[16]

---

[15]Kant, 204 (A 169, B 211).
[16]See du Bois-Reymond [1887, p. 66].

Points, lacking any sort of magnitude, can never melt and merge together into a magnitude of any size whatsoever, continuous or not. Peirce, by constructing his continuum from numbers, rather than points, does not avoid this criticism, but falls under it even more strongly, as numbers do not have a physical existence. At the least, a point can be viewed, and is viewed under du Bois-Reymond's Empiricist-built ontology, as a physical location upon a line and thus have a connection to the physical; individual numbers have no physical existence, and thus there simply is nothing which can merge with other numbers into an extended interval. Peirce would have done better to take continuity as a logical given, and used numbers, as he used points, to navigate upon this continuum, rather than attempting to form a continuum from supermultitudinously merging numbers.

I believe that Peirce's theory of continuity and the troubles contained within are the result of Peirce attempting to model this Aristotelian insight into continuity within a Dedekind-esque foundational framework. In so doing, Peirce has not only failed to avoid the many difficulties with compositional continua, but he has also conflated the property of continuity with the tools used to measure and analyze continuity, and in the process, paradoxically created a system that is entirely useless in mathematics. Peirce creates his continuity in much the same way as Dedekind creates his, by beginning with numbers and adding enough to reach continuity. Rather than stopping at the collection of real numbers as Dedekind did, however, Peirce believed more is necessary for true continuity, and continues to add more and more *numbers*.

By creating continuity from elements such as points or numbers, Peirce retained the philosophical problems that plagued Cantor and Dedekind's models, and this supermultitudinous collection is itself a compositional continuity by our definition above. By insisting that this set of discrete elements must be expanded until the elements themselves meld together, Peirce introduced a new host of difficulties. Without physical extension, these elements would never run out of room, and thus would never be forced to meld; furthermore, there is no physical extension with which to meld. If we were to do as Peirce directs, that is, take the set of real numbers and, between every two reals, insert the entire collection of real numbers between zero and one, the result would be equinumerous to the collection of real numbers itself, no matter how many times the process is repeated. The creation process thus fails miserably.

Assume, for a moment, that the creation of this supermultitudinous collection actually succeeded. By using points and/or numbers to create continuity, these tools necessary to measure parts of this continuity are the very things that are melded away into continuity. If numbers melt into one another in the process that forms a continuum, no method of quantifying parts of this continuum remains; if points are merged into a continuous entity, the points themselves are no longer available to us to mark distance on that entity. Peirce has, in fact, retained Cantor and Dedekind's conflation between the property of continuity and the elements used to measure continuous entities, and then he proceeded to destroy the ability of this collection to serve mathematical purposes.

Moreover, this defect cannot be repaired, given his theory. His main criticism of a real number continuum was that the real numbers are not the largest possible set, and true continuity must contain all possible elements. Thus, he cannot restore the mathematical viability of his continuum by introducing *new* points or numbers as tools of measurement. Were we to admit the existence of such points, we would also have to admit that the supermultitudinous collection did not contain all possible points to begin with.

Thus, Peirce accurately criticizes Cantor and Dedekind for attempting to create continuity from discrete elements, but ultimately falls into the same trap as they did. By attempting to overcome the flaws in their system while retaining a foundational mathematical approach to continuity, Peirce creates new problems and solves few.

8.6. Continuity and du Bois-Reymond

All that remains is to discuss the theory of continuity presented by du Bois-Reymond, under the assumptions of the Idealist. Unlike the Empiricist, the Idealist embraces abstract entities such as idealized shapes and lines, and draws logical conclusions from them. In doing so, he forms a distinct theory of continuity, one which applies specifically to the straight line but has consequences for number systems meant to measure this straight line. The key to his theory of continuity is contained in his argument for the existence of infinitesimals. The assumption on which the argument is based expresses a clear concept of continuity; if the line is continuous, then "the number of points of division of the unit length is infinitely large."[17] This clearly echoes Aristotle's idea of continuity; however, while Aristotle stated that continuous entities were divided into parts themselves divisible, it seems at first blush as though du Bois-Reymond's definition of continuity would require division into points – themselves non-divisible entities.

This is not the case, and in fact the next sentence of this proof clarifies the Idealist's stance. "According to the true concept of magnitude, these points do not follow each other without an interval." A continuous line segment, for the Idealist, is not divided into infinitely many points; rather, the segment is divided *by points* into infinitely many intervals. Points are the means of division, not the result of it, and these points must always maintain intervals between them. This theory of magnitude, and thus of continuity, is not a compositional one; this continuum is not composed of points. This is made most clear by the statement, "points alone can never form extensions."[18] If points cannot form extensions, they can never form a continuum. The commitment of the Idealist to an Aristotelian continuity is made clear when he argues, "every extension as small as it may be must be organized like the unit length, and similarly contain infinitely many points of division."[19] The Idealist further insists that if there are actually infinitely many points of division on a line segment, then these intervals must eventually become infinitesimal in length, but they never become *extensionless*; they are always intervals, which, one assumes, can themselves be divided.

It may be tempting at this point to claim that the problems which plagued the compositional theories of continua can be solved by composing our continuum from points and infinitesimal intervals; in other words, we may wish to argue that compositionality of continua is not itself a problem, but rather that points or numbers are particularly bad entities from which to compose continua. The Idealist's infinitesimal interval seems like a far better candidate for such a job, since composing a continuum from finite intervals would do nothing to explain the nature of continuity itself. Think, for example, of composing a long continuous line segment from shorter continuous line-segments; this would be in some sense compositional, but continuity would not be the property thereby

---

[17]See du Bois-Reymond [1887, p. 73].
[18]Ibid., p. 73.
[19]Ibid., p. 73.

constructed, it existed previously to our construction, and so would only be preserved.

However, composing a continuum from infinitesimal intervals does not solve all of our problems. First of all, if the infinitesimal interval is itself infinitely divisible, as Peirce's infinitesimals are, one could make the argument that these intervals were themselves continuous; indeed, they must be, if the mirror quality is to be respected. Thus, it is not continuity which is composed from infinitesimals, but rather, a larger continuity would be built from a collection of smaller continuous elements. That du Bois-Reymond's Idealist believes these infinitesimal intervals are at least themselves infinitely divisible is evident from his assertion that "every extension as small as it may be must be organized like the unit length, and similarly contain infinitely many points of division." This clearly is meant to include infinitesimal extensions as well as finite ones.

Another difficulty plagues the proposed composition of continuity from infinitesimal magnitudes; such a construction would commit the same conflation error seen in Dedekind, Cantor, and Peirce. Infinitesimal magnitudes are, like the unit interval, artifacts of measuring and quantifying, and there is no reason to suppose that they are inherently contained in all continuous entities, assuming any such entities actually exist. Composing a continuum from infinitesimals and points is no better than composing a continuum from points alone. As an analogy, consider the unit segment; let us specify the unit segment as a line segment of precisely an inch in length. Because one could theoretically construct an infinite line from an infinite number of inch-long line segments is no reason to suppose that the inch itself is an inherent component of any line, and *a fortiori*, that the inch is an inherent component of any continuum.

It must be noted, however, that du Bois-Reymond's Idealist does not make this mistake; he does not speak in terms of *composing* a continuum from infinitesimal intervals. Rather, the conclusion of his argument is "the unit extension is decomposed into an infinity of partial extensions, of which none is finite. Thus the infinitely small really exists."[20] The infinitesimal in the Idealist's system is not a necessary element of the composition of a continuum. Rather, it is a necessary by-product of decomposing a continuous extension. An infinitesimal interval results when a finite continuous extension is divided into infinitely

---

[20]Ibid., p. 73.

many parts, as "before the parts of a finite quantity can be infinite in number, each must be infinitely small."[21] Du Bois-Reymond's argument for the existence of infinitesimals is not that they are a necessary element of continuity, but rather that they are a necessary result from a continuous entity which is finite in extension but infinitely divided.

It is possible that du Bois-Reymond's Idealist does conflate the property of continuity with the tools for measuring continuous elements; one piece of evidence for such a conflation is the seeming interchangeability of the concepts of point and number in du Bois-Reymond's writing. This is not necessarily a conclusive indication of this conflation, however, as points and numbers are themselves both ways of quantifying and measuring magnitude. His non-Archimedean continuity, which can be decomposed into points and intervals, is an interesting one. Of the four theories of continuity examined in this book, it is the one which is the most compatible with the Aristotelian and Kantian insight that any part into which a continuum is divided must itself exhibit the basic properties of continuity.

8.7. Conclusion

In this last chapter I have argued that a true understanding of continuity must make a sharp distinction between the property of continuity itself and the tools for measuring continuity. Though tools of analysis are necessary in one sense to comprehend and manipulate continuous entities (numbers, for example, are extremely useful for measuring the passage of time), one must not confuse comprehending a continuum in this manner with comprehending the essential features of continuity itself. Cantor, Dedekind, and Peirce all conflate the property with our tools of measurement, and all run afoul of philosophical difficulties as a result. Du Bois-Reymond's Idealist system has philosophical difficulties of its own, which even du Bois-Reymond himself admitted (for example, it is wholly divorced from any empirical experience of the world), but by avoiding the category error of the other systems, it is the most promising of the four developments of the concept of continuity. Rather than simply ending with a summary of my main argument, however, I would like to end by permitting

---

[21]Ibid., p. 83.

myself some remarks on three larger issues, which are strictly speaking outside of the range of the current project, and yet, worth at least noting.

First, continuity itself. Notice that while I have insisted that continua are not composed of the things we use to measure continuous entities, but rather that continuity is a property entirely distinct from mathematics, I have not further specified any particular features of continuity. I do believe that Aristotle's insight, that continuous things must be divisible into parts which themselves are divisible, holds the key to continuity; however, beyond that, continuity remains something of a mystery. Du Bois-Reymond's Empiricist very well may be on the right track when he claims that continuity is beyond our direct experience. However, it seems more likely to me that a Kantian evaluation would be more accurate; that continuity, like space and time itself, is a necessary feature of the way in which we experience the world. Whatever the true nature of continuity itself, it seems obvious that if such a thing exists, in space, or time, or elsewhere, it is a property quite independent of our means of quantifying and analyzing continuous things.

As for infinitesimals, they should not be rejected as useless, as they could contribute much of value to mathematics. I have argued to this effect in Chapter 7. However, while they are not useless, neither are they necessary elements of continuous systems. Infinitesimal quantities, like a variety of mathematical systems and entities, are tools with very specific uses. It may well be that infinitesimals are indispensable tools for the most comprehensive analysis of continuous entities, but just as continua are not dependent upon numbers or points, neither are they dependent upon infinitesimally small magnitudes. A non-standard analysis, such as the one presented by Abraham Robinson, may well turn out to be the system of calculus that is maximally useful for thorough analysis of continuous phenomenon, but even this would not prove infinitesimal quantities were necessary elements of these continuous phenomenon themselves.

Finally, Cantor and Dedekind were mistaken about the metaphysical importance of their mathematical systems; they both failed to produce necessary and sufficient conditions for defining continuous phenomenon. However, their mathematical systems are far from useless. By providing a system of numbers which guarantees that wherever one cuts the number system, one does so at a point, Dedekind gave mathematicians a system of real numbers which was both defined independently of geometry, and uniquely suited to be wed back to geometry, with the use of the Cantor-Dedekind axiom. By providing necessary and sufficient conditions for a "continuous" real number system, Cantor did the same – he too created a system of real numbers which is uniquely suited to be used to quantify continuous things. Rather than creating continua from numbers, Cantor and Dedekind accomplished something much more mathematically interesting and important – they created systems of real numbers which guarantee that no matter how fine-grained we get in our analysis of continuous phenomenon, our numbers will not fail us. By using a Cantorian or Dedekindesque system of real numbers, we are guaranteed that we can measure most phenomena with as much accuracy as we wish without running into the incommensurability problems which plagued mathematics before the sixteenth century. We are also guaranteed that mathematics can stand on arithmetic feet, without constant reference to geometry – which was, after all, Dedekind's original goal in creating his real number system.

This does not mean that Cantor or Dedekind provided the one and only system of numbers appropriate for analysis of continuous entities. A Cantor/Dedekind system of real numbers, and a non-standard system of numbers, can both be useful tools in our quest to understand the wider world. Both should be developed, as Cantor recommends, without constant reference to applied mathematics, without constant reference to precisely what they are best at measuring, as such a creative approach to mathematics is not only interesting in itself, but provides the intellectual community with a variety of mathematical tools, some of which may be better suited than others to particular jobs.

Bibliography

Francine F. Abeles. The enigma of the infinitesimal: toward Charles L. Dodgson's theory of infinitesimals. *Modern Logic*, 8(3-4):7–19, 2000.

Jonathan Barnes. *The Complete Works of Aristotle*. Princeton University Press, Princeton, 1984.

John L. Bell. Dissenting voices: Divergent conceptions of the continuum in 19th and early 20th century mathematics and philosophy. [Online]publish.uwo.ca/\%7Ejbell/Dissenting\%20Voice1.pdf.

John L. Bell. Continuity and infinitesimals. In Edward N. Zalta, editor, *The Stanford Encyclopedia of Philosophy*. Stanford University, Metaphysics Research Lab, Stanford, Fall 2005. [Online]plato.stanford.edu/archives/fall2005/entries/continuity/.

George Berkeley. *The Analyst*; or, a discourse addressed to an infidel mathematician. In Alexander Campbell Fraser, editor, *Works of George Berkeley*. Clarendon Press, Oxford, 1901. *The Analyst* was originally published in 1734.

Carl B Boyer. *The History of the Calculus and its Conceptual Development*. Dover Publications, New York, 1959. Originally published by Hafner in 1949.

Joseph Brent. *Charles Sanders Peirce: a Life*. Indiana University Press, Bloomington, 1993.

Robert Burch. Charles Sanders Peirce. In Edward N. Zalta, editor, *The Stanford Encyclopedia of Philosophy*. Stanford University, Metaphysics Research Lab, Stanford, Fall 2006. [Online]plato.stanford.edu/archives/fall2006/entries/peirce/.

Georg Cantor. Beweis, dass eine für jeden reellen Wert von x durch eine trigonometrische Reihe gegebene Funktion $f(x)$ sich nur auf eine einzige Weise in dieser Form darstellen lässt. *Journal für die reine und angewandte Mathematik*, 72:139–142, 1870.

Georg Cantor. Über die Ausdehnung eins Satzes aus der Theorie der trigonometrischen Reihen. *Mathematische Annalen*, 5:123–132, 1872. French translation published as "Extension d'un théorem de la theorie des series trigonometriques." *Acta Mathematica* 2 (1883): 336–348.

Georg Cantor. Ein Beitrag zur Mannigfaltigkeitslehre. *Journal für die rein und angewandte Mathematik*, 94:242–258, 1878. French translation published as "Une contribution à la théorie des ensembles." *Acta Mathematica* 2 (1878) 311–328.

Georg Cantor. *Grundlagen einer allgemeinen Mannigfaltigkeitslehre. Ein mathematischphilosophischer Versuch in der Lehre des Unendlichen.* B. G. Teübner, Leipzig, 1883.

Georg Cantor. *Contributions to the Founding of the Theory of Transfinite Numbers.* Dover Publications, New York, 1955. Translated by Philip E. B. Jourdain.

Georg Cantor. *Gesammelte Abhandlungen mathematischen und philosophischen Inhalts.* Olms, Hildesheim, 1966. E. Zermelo (ed.).

Joseph W. Dauben. C. S. Peirce's philosophy of infinite sets: a study of Peirce's interest in the infinite related to the birth of American mathematics and contemporary work of Cantor and Dedekind. *Mathematics Magazine*, 50(3): 123–135, May 1977.

Joseph W. Dauben. *Georg Cantor: His Mathematics and Philosophy of the Infinite.* Princeton University Press, Princeton, 1990.

Richard Dedekind. *Essays on the Theory of Numbers.* Dover Publications, New York, 1963. This small volume contains translations of both *Stetigkeit und irrationale Zahlen* and *Was sind und was sollen die Zahlen?*

Richard Dedekind. *Richard Dedekind, 1831-1981: Eine Würdigung zu seinem 150 Geburtstag.* Vieweg, Braunschweig, 1981.

Rene Descartes. *Meditations on First Philosophy.* Cambridge University Press, Cambridge, 1996. Originally published in 1644.

Paul du Bois-Reymond. *De aequilibrio fluidorum.* Ph.D., University of Berlin, 1859. [Online]gdz.sub.unigoettingen.de/dms/load/img/?IDDOC=42460.

Paul du Bois-Reymond. Sur la grandeur relative des infinis des fonctions. *Annali di matematica pura de applicata*, 4:338–353, 1870. Series IIa.

Paul du Bois-Reymond. Ueber die Paradoxen des Infinitärcalcüls. *Mathematische Annalen*, 11:149–167, 7 1877.

Paul du Bois-Reymond. *Théorie générale des fonctions.* Imprimerie Niçoise, Nice, 1887. Translated by Gaston Milhaud. French translation of *Die allgemeine Functiontheorie.* Du Bois-Reymond wrote a preface to this translation, in which he claims to have personally gone over this translation, and fixed some things he considered to be errors or not well explained in the original

German.

William Dunham. *Journey Through Genius: The Great Theorems of Mathematics*. Penguin Books, New York, 1991.

William Dunham. *Euler: The Master of Us All*. The Mathematical Association of America, Washington D.C., 1999.

Philip Ehrlich, editor. *Real Numbers, Generalizations of the Reals, and Theories of Continua*. Kluwer Academic Publishers, Dordrecht, 1994.

Euclid. *The Thirteen Books of The Elements, Vol. I*. Dover Publications, New York, 1956. Translated by Thomas L. Heath.

Leonhard Euler. *Introduction to Analysis of the Infinite, Book 1*. Springer-Verlag, New York, 1988. Translated by John Blanton.

James K. Feibleman. *An Introduction to Peirce's Philosophy*. Hauser Press, New Orleans, 1946.

José Ferreriós. On the relations between Georg Cantor and Richard Dedekind. *Historia Mathematica*, 20:343–363, 1993.

Gordon Fisher. The infinite and infinitesimal quantities of du Bois-Reymond and their reception. *Archive for History of Exact Sciences*, 24:101–164, 1981.

Ivor Grattan-Guinness. *The Search for Mathematical Roots, 1870-1940: Logics, Set Theories, and the Foundations of Mathematics from Cantor through Russell to Gödel,*. Princeton University Press, Princeton, 2001.

Spencer Gwartney-Gibbs. Continuous frustration: C. S. Peirce's mathematical conception of continuity. Presented at the 2007 meeting of the Society for the Advancement of American Philosophy, 2007. [Online]www.philosophy.uncc.edu/mleldrid/SAAP/USC/DP16.html.

G. H. Hardy. *Orders of Infinity: the Infinitärcalcül of Paul du Bois-Reymond*. Cambridge University Press, Cambridge, 1924.

Carl Hausman. Infinitesimals as origins of evolution: Comments prompted by Timothy Herron and Hilary Putnam on Peirce's synechism and infinitesimals. *Transactions of the Charles S. Peirce Society*, 34(3):627–640, Summer 1998.

Timothy Herron. C. S. Peirce's theories of infinitesimals. *Transactions of the Charles S. Peirce Society*, 23(3):590–645, Summer 1997.

Edward V. Huntington. *The Continuum and Other Types of Serial Order: with an introduction to Cantor's transfinite numbers*. Dover Press, Mineola, 1917. [Online]www.histanalytic.org/Huntington.htm.

Immanuel Kant. *The Critique of Pure Reason*. St. Martin's Press, New York, 1965. Translated by Norman Kemp Smith. Reference to this work will be by standard A-deduction/B-deduction pagination (for example, A26/B42).

Norman Kretzmann. Incipit/desinit. In Peter K. Machamer and Robert G. Turnbull, editors, *Motion and Time, Space and Matter: Interrelations in the History of Philosophy and Science*. Ohio State University Press, Columbus, 1976.

Gottfried Wilhelm Leibniz. *The Labyrinth of the Continuum: Writings on the Continuum Problem, 1672–1686*. Yale University Press, New Haven, 2001. Edited and translated by Richard T. W. Arthur.

Stephen H. Levy. Charles S. Peirce's theory of infinitesimals. *International Philosophical Quarterly*, 31:127–140, 1991.

David Charles McCarty. David Hilbert and Paul du Bois-Reymond: Limits and ideals. In G. Link, editor, *One Hundred Years of Russell's Paradox*, pages 517–532. De Gruyter, Berlin, 2004.

Matthew E. Moore. A Cantorian argument against infinitesimals. *Synthese*, 133:305–330, 2002.

Matthew E. Moore. Peirce's Cantor. In Matthew E. Moore, editor, *New Essays on Peirce's Mathematical Philosophy*, pages 323–362. Open Court, Chicago, 2010.

William of Ockham. *William of Ockhams Quodlibetal Questions*. Yale University Press, New Haven, 1991. Translated by Alfred J. Freddoso and Francis E. Kelley.

U. Parpart. Foundations of the theory of manifolds. *The Campaigner*, 9:69–96, 1976.

Charles Sanders Peirce. The law of mind. *The Monist*, 2(3):533–559, July 1892.

Charles Sanders Peirce. Evolutionary love. *The Monist*, 3(1):176–200, 1893.

Charles Sanders Peirce. *Collected Papers of Charles Sanders Peirce. Volumes I-VIII*. Belknap Press of Harvard University Press, Cambridge, 1960. Edited by Charles Hartshorn and Paul Weiss.

Charles Sanders Peirce. *The New Elements of Mathematics*. Mouton Publishers, The Hague, 1976. Edited by Carolyn Eisele.

Charles Sanders Peirce, editor. *The Essential Peirce: Selected Philosophical Writings*. Indiana University Press, Bloomington, 1992. Nathan Houser (ed.).

Charles Sanders Peirce. *The Essential Peirce: Selected Philosophical Writings, Volume 2 (1893–1913)*. Indiana University Press, Bloomington, 1998a. Edited by The Peirce Edition Project.

Charles Sanders Peirce. *Chance, Love, and Logic*. University of Nebraska Press, Lincoln, 1998b. Morris Cohen (ed.).

Plato. *Five Dialogues: Euthyphro, Apology, Crito, Meno, Phaedo*. Hackett Publishing Company, Indianapolis, 2002. Translated by G. M. A. Grube.

Vincent G. Potter and Paul B. Shields. Peirce's definitions of continuity. *Transactions of the Charles S. Peirce Society*, 13:20–34, Winter 1977.

Abraham Robinson. *Non-Standard Analysis*. Princeton University Press, Princeton, 1996.

Bertrand Russell. *The Principles of Mathematics*. George Allen & Unwin Ltd., London, 1903.

David Eugene Smith. *A Source Book in Mathematics*. McGraw Hill Book Company, New York, 1929.

Paul Spade. How to start and stop: Walter Burley on the instant of transition. *The Journal of Philosophical Research*, 19:193–221, 1994.

Baruch Spinoza. *Ethics*. Prometheus Books, Amherst, 1989. Translated by R. H. M. Ewles.

Edith Dudley Sylla. The Oxford calculators. In Norman Kretzmann, Anthony Kenny, and Jan Pinborg, editors, *The Cambridge History of Later Medieval Philosophy*. Cambridge University Press, Cambridge, 1982.

Friedrich Waismann. *Introduction to Mathematical Thinking: the formation of concepts in modern mathematics*. Dover Publications, Mineola, 2003. Translated by Theodore J. Benac.

Guillermina Waldegg. Bolzano's approach to the paradoxes of infinity: Implications for teaching. *Science & Education*, 14(6):559–577, Aug 2005.

Philip Wiener, editor. *Charles S. Peirce: Selected Writings (Values in a Universe of Chance)*. Dover, New York, 1966.

Printed in Germany
by Amazon Distribution
GmbH, Leipzig